Encyclopaedia of

Genes, Genomes, Genetics and Chromosomes

Encyclopaedia of Genes, Genomes, Genetics and Chromosomes

Volume 2

Editor:
Dr. R.N. Shukla

AGROTECH PRESS
JAIPUR ▪ NEW DELHI

AGROTECH PRESS

Regd. Off. "Bony Residency", Gate No. 2
Opp. Tilak Public School, Vishweswariya Nagar,
Gopalpura Road, Jaipur - 302018, India
Ph: 0141 - 2761280, Telefax: 2761381
E-mail : agrotechpress@gmail.com

Delhi Office: 102, 1st Floor, Satyam House,
4327/3, Ansari Road, Darya Ganj,
New Delhi - 110002, India
Ph: 011-45652440
visit us: www.agrotechpress.com

Distributed By
KEYA INTERNATIONAL PRESS LLP
E-mail: keyainternationalpressllp@gmail.com

Published in India, 2024

ISBN : 9789384568238 (5 Vols. Set)

Typeset by Thomson Press

Printed and bound in India by Thomson Press Ltd.,
New Delhi

Contents

Preface

In modern molecular biology and genetics, the genome is the entirety of an organism's hereditary information. It is encoded either in DNA or, for many types of virus, in RNA. The genome includes both the genes and the non-coding sequences of the DNA/RNA. The term was adapted in 1920 by Hans Winkler, professor of botany at the University of Hamburg, Germany. The Oxford English Dictionary suggests the name to be a blend of the words *gene* and *chromosome*. A few related *-ome* words already existed, such as *biome* and *rhizome*, forming a vocabulary into which *genome* fits systematically. In 1976, Walter Fiers at the University of Ghent was the first to establish the complete nucleotide sequence of a viral RNA-genome. The next year, Phage Ö-X174, with only 5386 base pairs, became the first DNA-genome project to be completed, by Fred Sanger. The first complete genome sequences for representatives from all 3 domains of life were released within a short period during the mid-1990s. The first bacterial genome to be sequenced was that of Haemophilus influenzae, completed by a team at The Institute for Genomic Research in 1995. A few months later, the first eukaryotic genome was completed, with the 16 chromosomes of budding yeast Saccharomyces cerevisiae being released as the result of a European-led effort begun in the mid-1980s. Shortly afterwards, in 1996, the first genome sequence for an archaeon, Methanococcus jannaschii, was completed, again by The Institute for Genomic Research. The development of new technologies has made it dramatically easier and cheaper to do sequencing, and the number of complete genome sequences is growing rapidly. The US National Institutes of Health maintains one of several comprehensive databases of genomic information. Among the thousands of completed genome sequencing projects include those for mouse, rice, the plant *Arabidopsis thaliana*, the puffer fish, and bacteria like E. coli. New sequencing technologies, such as massive parallel sequencing have also opened up the prospect of personal genome sequencing as an important diagnostic tool, as pioneered by Manteia Predictive Medicine. A major step towards that goal was the completion in 2007 of the full genome of James D. Watson, one of the co-discoverers of the structure of DNA. Whereas a genome sequence lists the order of every DNA base in a genome, a genome map identifies the landmarks. A genome map is less detailed than a genome sequence and aids in navigating around the genome. The Human Genome Project was organized to map and to sequence the human genome. A fundamental step in the project was the release of a detailed genomic map by Jean Weissenbach and his team at the Genoscope in Paris. Genomes are more than the sum of an organism's genes and have traits that may be measured and studied without reference to the details of any particular genes and their products. Researchers compare traits such as *chromosome number*, genome size, gene order, codon usage bias, and GC-content to determine what mechanisms could have produced the great variety of genomes that exist today Duplications play a major role in shaping the genome. Duplications may range from extension of short tandem repeats, to duplication of a cluster of genes, and all the way to duplications of entire chromosomes or even entire genomes. Such duplications are probably fundamental to the creation of genetic novelty. Horizontal gene transfer is invoked to explain how there is often extreme similarity between small portions of the genomes of two organisms that are otherwise very distantly related. Horizontal gene transfer seems to be common among many microbes. Also, eukaryotic cells seem to have experienced a transfer of some genetic material from their chloroplast and mitochondrial genomes to their nuclear chromosomes. The concept of the gene has changed considerably. From the original definition of a "unit of inheritance", the term evolved to mean a DNA-based unit that can exert its effects on the organism through RNA or protein products. It was also previously believed that one gene makes one protein; this concept was overthrown by the discovery of alternative splicing and trans-splicing.

The present publication is an extensive work on genome and covers much more of the subject as is generally expected of students going up for the examinations of several Universities in India and abroad. In spite of the comprehensiveness, the subject matter has been presented in such a way that it is easily accessible to the students.

—Editor

1

Genetic Code

The genetic code is the set of rules by which information encoded in genetic material (DNA or mRNA sequences) is translated into proteins (amino acid sequences) by living cells. The code defines a mapping between tri-nucleotide sequences, called codons, and amino acids. With some exceptions, a triplet codon in a nucleic acid sequence specifies a single amino acid. Because the vast majority of genes are encoded with exactly the same code, this particular code is often referred to as the canonical or standard genetic code, or simply *the* genetic code, though in fact there are many variant codes. For example, protein synthesis in human mitochondria relies on a genetic code that differs from the standard genetic code.

Not all genetic information is stored using the genetic code. All organisms' DNA contains regulatory sequences, intergenic segments, and chromosomal structural areas that can contribute greatly to phenotype. Those elements operate under sets of rules that are distinct from the codon-to-amino acid paradigm underlying the genetic code.

Discovery

After the structure of DNA was deciphered by James Watson, Francis Crick, Maurice Wilkins and Rosalind Franklin, serious efforts to understand the nature of the encoding of proteins began. George Gamow postulated that a three-letter code must be employed to encode the 20 standard amino acids used by living cells to encode proteins, because 3 is the smallest integer n such that 4 is at least 20.

The fact that codons consist of three DNA bases was first demonstrated in the Crick, Brenner et al. experiment. The first elucidation of a codon was done by Marshall Nirenberg and Heinrich J. Matthaei in 1961 at the National Institutes of Health. They used a cell-free system to translate a poly-uracil RNA sequence (i.e., UUUUU...) and discovered that the polypeptide that they had synthesized consisted of only the amino acid phenylalanine. They thereby deduced that the codon UUU specified the amino acid phenylalanine.

This was followed by experiments in the laboratory of Severo Ochoa demonstrating that the poly-adenine RNA sequence (AAAAA...) coded for the polypeptide, poly-lysine. and the poly-cytosine RNA sequence (CCCCC...) coded for the polypeptide, poly-proline. Therefore the codon AAA specified the amino acid lysine, and the codon CCC specified the amino acid proline.

Using different copolymers most of the remaining codons were then determined. Extending this work, Nirenberg and Philip Leder revealed the triplet nature of the genetic code and allowed the codons of the standard genetic code to be deciphered. In these experiments various combinations of mRNA were passed through a filter which contained ribosomes, the components of cells that translate RNA into protein. Unique triplets promoted the binding of specific tRNAs to the ribosome. Leder and Nirenberg were able to determine the sequences of 54 out of 64 codons in their experiments.

Subsequent work by Har Gobind Khorana identified the rest of the genetic code. Shortly thereafter, Robert W. Holley determined the structure of transfer RNA (tRNA), the adapter molecule that facilitates the process of translating RNA into protein. This work was based upon earlier studies by Severo Ochoa, who received the Nobel prize in 1959 for his work on the enzymology of RNA synthesis. In 1968, Khorana, Holley and Nirenberg received the Nobel Prize in Physiology or Medicine for their work.

Transfer of Information Via the Genetic Code

The genome of an organism is inscribed in DNA, or in the case of some viruses, RNA. The portion of the genome that codes for a protein or an RNA is referred to as a gene. Those genes that code for proteins are composed of tri-nucleotide units called codons, each coding for a single amino acid. Each nucleotide sub-unit consists of a phosphate, deoxyribose sugar and one of the 4 nitrogenous nucleobases. The purine bases adenine (A) and guanine (G) are larger and consist of two aromatic rings. The pyrimidine bases cytosine (C) and thymine (T) are smaller and consist of only one aromatic ring. In the double-helix configuration, two strands of DNA are joined to each other by hydrogen bonds in an arrangement known as base pairing. These bonds almost always form between an adenine base on one strand and a thymine on the other strand and between a cytosine base on one strand and a guanine base on the other. This means that the number of A and T residues will be the same in a given double helix, as will the number of G and C residues. In RNA, thymine (T) is replaced by uracil (U), and the deoxyribose is substituted by ribose.

Each protein-coding gene is transcribed into a template molecule of the related polymer RNA, known as messenger RNA or mRNA. This, in turn, is translated on the ribosome into an amino acid chain or polypeptide. The process of translation requires transfer RNAs specific for individual amino acids with the amino acids covalently attached to them, guanosine triphosphate as an energy source, and a number of translation factors.

tRNAs have anticodons complementary to the codons in mRNA and can be "charged" covalently with amino acids at their 3' terminal CCA ends. Individual tRNAs are charged with specific amino acids by enzymes known as aminoacyl tRNA synthetases, which have high specificity for both their cognate amino acids and tRNAs. The high specificity of these enzymes is a major reason why the fidelity of protein translation is maintained.

There are $4^3 = 64$ different codon combinations possible with a triplet codon of three nucleotides; all 64 codons are assigned for either amino acids or stop signals during translation. If, for example, an RNA sequence, UUUAAACCC is considered and the reading frame starts with the first U (by convention, 5' to 3'), there are three codons, namely, UUU, AAA and CCC, each of which specifies one amino acid. This RNA sequence will be translated into an amino acid sequence, three amino acids long. A comparison may be made with computer science, where the codon is similar to a word, which is the standard "chunk" for handling data (like one amino acid of a protein), and a nucleotide is similar to a bit, in that it is the smallest unit.

Salient Features

Sequence Reading Frame

A codon is defined by the initial nucleotide from which translation starts. For example, the string GGGAAACCC, if read from the first position, contains the codons GGG, AAA and CCC; and, if read from the second position, it contains the codons GGA and AAC; if read starting from the third position, GAA and ACC. Every sequence can thus be read in three reading frames, each of which will produce a different amino acid sequence (in the given example, Gly-Lys-Pro, Gly-Asn, or Glu-Thr, respectively). With double-stranded DNA there are six possible reading frames, three in the forward orientation on one strand and three reverse on the opposite strand. The actual frame in which a protein sequence is translated is defined by a start codon, usually the first AUG codon in the mRNA sequence.

Start/Stop Codons

Translation starts with a chain initiation codon (start codon). Unlike stop codons, the codon alone is not sufficient to begin the process. Nearby sequences (such as the Shine-Dalgarno sequence in *E. coli*) and initiation factors are also required to start translation. The most common start codon is AUG which is read as methionine or, in bacteria, as formylmethionine. Alternative start codons (depending on the organism), include "GUG" or "UUG", which normally code for valine or leucine, respectively. However, when used as a start codon, these alternative start codons are translated as methionine or formylmethionine.

The three stop codons have been given names: UAG is *amber*, UGA is *opal* (sometimes also called *umber*), and UAA is *ochre*. "Amber" was named by discoverers Richard Epstein and Charles Steinberg after their friend Harris Bernstein, whose last name means "amber" in German. The other two stop codons were named "ochre" and "opal" in order to keep the "color names" theme. Stop codons are also called "termination" or "nonsense" codons and they signal release of the nascent polypeptide from the ribosome due to binding of release factors in the absence of cognate tRNAs with anticodons complementary to these stop signals.

Effect of Mutations

During the process of DNA replication, errors occasionally occur in the polymerization of the second strand. These errors, called mutations, can have an impact on the phenotype of an organism, especially if they occur within the protein coding sequence of a gene. Error rates are usually very low—1 error in every 10–100 million bases—due to the "proofreading" ability of DNA polymerases.

Missense mutations and nonsense mutations are examples of point mutations, which can cause genetic diseases such as sickle-cell disease and thalassemia respectively. Clinically important missense mutations generally change the properties of the coded amino acid residue between being basic, acidic polar or non-polar, whereas nonsense mutations result in a stop codon.

Mutations that disrupt the reading frame sequence by indels (insertions or deletions) of a non-multiple of 3 nucleotide bases are known as frameshift mutations. These mutations usually result in a completely different translation from the original, and are also very likely to cause a stop codon to be read, which truncates the creation of the protein. These mutations may impair the function of the resulting protein, and are thus rare in *in vivo* protein-coding sequences. One reason inheritance of frameshift mutations is rare is that if the protein being translated is essential for growth under the selective pressures the organism faces, absence of a functional protein may cause death before the organism is viable. Frameshift mutations may result in severe genetic diseases such as Tay-Sachs disease. Although most mutations that change protein sequences are harmful or neutral, some mutations have a positive effect on an organism. These mutations may enable the mutant organism to withstand particular environmental stresses better than wild-type organisms, or reproduce more quickly.

In these cases a mutation will tend to become more common in a population through natural selection. Viruses that use RNA as their genetic material have rapid mutation rates, which can be an advantage since these viruses will evolve constantly and rapidly, and thus evade the defensive responses of e.g. the human immune system. In large populations of asexually reproducing organisms, for example, E. coli, multiple beneficial mutations may co-occur, causing competition among them, this phenomenon is called clonal interference.

Degeneracy of the Genetic Code

The genetic code has redundancy but no ambiguity. For example, although codons GAA and GAG both specify glutamic acid (redundancy), neither of them specifies any other amino acid (no ambiguity). The codons encoding one amino acid may differ in any of their three positions. For example the amino acid glutamic acid is specified by GAA and GAG codons (difference in the third position), the amino acid leucine is specified by UUA, UUG, CUU, CUC, CUA, CUG codons (difference in the first or third position), while the amino acid serine is specified by UCA, UCG, UCC, UCU, AGU, AGC (difference in the first, second or third position).

A position of a codon is said to be a fourfold degenerate site if any nucleotide at this position specifies the same amino acid. For example, the third position of the glycine codons (GGA, GGG, GGC, GGU) is a fourfold degenerate site, because all nucleotide substitutions at this site are synonymous; i.e., they do not change the amino acid. Only the third positions of some codons may be fourfold degenerate.

A position of a codon is said to be a twofold degenerate site if only two of four possible nucleotides at this position specify the same amino acid. For example, the third position of the glutamic acid codons (GAA, GAG) is a twofold degenerate site. In twofold degenerate sites, the equivalent nucleotides are always either two purines (A/G) or two pyrimidines (C/U), so only transversional substitutions (purine to pyrimidine or pyrimidine to purine) in twofold degenerate sites are nonsynonymous. A position of a codon is said to be a non-degenerate site if any mutation at this position results in amino acid substitution. There is only one threefold degenerate site where changing to three of the four nucleotides may have no effect on the amino acid (depending on what it is changed to), while changing to the fourth possible nucleotide always results in an amino acid substitution. This is the third position of an isoleucine codon: AUU, AUC, or AUA all encode isoleucine, but AUG encodes methionine. In computation this position is often treated as a twofold degenerate site.

There are three amino acids encoded by six different codons: serine, leucine, and arginine. Only two amino acids are specified by a single codon. One of these is the amino-acid methionine, specified by the codon AUG, which also specifies the start of translation; the other is tryptophan, specified by the codon UGG. The degeneracy of the genetic code is what accounts for the existence of synonymous mutations.

Degeneracy results because a triplet code designates 20 amino acids and a stop codon. Because there are four bases, triplet codons are required to produce at least 21 different codes. For example, if there were two bases per codon, then only 16 amino acids could be coded for (4^2=16). Because at least 21 codes are required, then 4^3 gives 64 possible codons, meaning that some degeneracy must exist.

These properties of the genetic code make it more fault-tolerant for point mutations. For example, in theory, fourfold degenerate codons can tolerate any point mutation at the third position, although codon usage bias restricts this in practice in many organisms; twofold degenerate codons can tolerate one out of the three possible point mutations at the third position. Since transition mutations (purine to purine or pyrimidine to pyrimidine mutations) are more likely than transversion (purine to pyrimidine or vice-versa) mutations, the equivalence of purines or that of pyrimidines at twofold degenerate sites adds a further fault-tolerance.

A practical consequence of redundancy is that some errors in the genetic code only cause a silent mutation or an error that would not affect the protein because the hydrophilicity or hydrophobicity is maintained by equivalent substitution of amino acids; for example, a codon of NUN (where N = any nucleotide) tends to code for hydrophobic amino acids. NCN yields amino acid residues that are small in size and moderate in hydropathy; NAN encodes average size hydrophilic residues. These tendencies may result from the shared ancestry of the aminoacyl tRNA synthetases related to these codons.

Even so, single point mutations can still cause dysfunctional proteins. For example, a mutated hemoglobin gene causes sickle-cell disease. In the mutant hemoglobin a hydrophilic glutamate (Glu) is substituted by the hydrophobic valine (Val), that is, GAA or GAG becomes GUA or GUG. The substitution of glutamate by valine reduces the solubility of â-globin which causes hemoglobin to form linear polymers linked by the hydrophobic interaction between the valine groups causing sickle-cell deformation of erythrocytes. Sickle-cell disease is generally not caused by a *de novo* mutation. Rather it is selected for in malarial regions (in a way similar to thalassemia), as heterozygous people have some resistance to the malarial *Plasmodium* parasite (heterozygote advantage).

These variable codes for amino acids are allowed because of modified bases in the first base of the anticodon of the tRNA, and the base-pair formed is called a wobble base pair. The modified bases include inosine and the Non-Watson-Crick U-G basepair.

Variations to the Standard Genetic Code

While slight variations on the standard code had been predicted earlier, none were discovered until 1979, when researchers studying human mitochondrial genes discovered they used an alternative code. Many slight variants have been discovered since, including various alternative mitochondrial codes, as well as small variants such as *Mycoplasma* translating the codon UGA as tryptophan and *Candida* species translating CUG as a serine rather than a leucine. In bacteria and archaea, GUG and UUG are common start codons. However, in rare cases, certain specific proteins may use alternative initiation (start) codons not normally used by that species.

In certain proteins, non-standard amino acids are substituted for standard stop codons, depending upon associated signal sequences in the messenger RNA: UGA can code for selenocysteine and UAG can code for pyrrolysine as discussed in the relevant articles. Selenocysteine is now viewed as the 21st amino acid, and pyrrolysine is viewed as the 22nd.

Notwithstanding these differences, all known codes have strong similarities to each other, and the coding mechanism is the same for all organisms: three-base codons, tRNA, ribosomes, reading the code in the same direction and translating the code three letters at a time into sequences of amino acids.

Expanded Genetic Code

An expanded genetic code refers to an artificially modified genetic code in which one or more specific codons have been allocated to encode an amino acid which is not among the twenty/twentytwo found in nature.

In order to understand how expansion of the genetic code is achieved, the translational mechanism catalysed by ribosomes should be considered: the central parts are the transfer RNAs (tRNA) which are used as keys to decode a three nucleotide code on the RNA being translated into its equivalent amino acid. To do this the tRNA recognises a specific codon thanks to a complementary sequence called the anticodon on one of its loops.

As a result generally there are one (or more) tRNAs for each codon (or group of degenerate codons). The encoding of a codon to its amino acid is a result of the aminoacyl tRNA synthetase which adds the aminoacyl group to its allocated tRNA. However, the aminoacyl tRNA synthetase often recognises a specific motif irrespective of the anticodon, meaning that if the anticodon were to be mutated the encoding of that amino acid would change to a new codon.

For successful translation of a novel amino acid, the codon to which the amino acid is reassigned must be free or unfavoured and the novel tRNA and synthetase set (called the orthogonal set when including the codon) must not crosstalk with the endogenous tRNA and synthetase sets, while still being functionally compatible with the ribosome and other components of the translation apparatus. The tRNA synthetase pair is taken from a distant organism, generally from a different domain, and the active site of the synthetase is modified to accept the non-natural amino acid.

The possibility of reassigning codons was realized by Normanly et al. in 1990 when a viable mutant strain of E. coli read through the amber codon. As a result the amber codon became the choice codon to be assigned a novel amino acid. Later, in the Schultz lab the tRNATyr/tyrosyl-tRNA synthetase (TyrRS) from Methanococcus jannaschii was used to introduce a tyrosine instead of STOP, the default value of the amber codon. As mentioned, this was possible because of the differences between the endogenous bacterial synthases and the orthologous archeal synthase which do not recognise each other.

Directed Evolution

Several methods for selecting the synthetase which accepts only the non natural amino acid have been developed. One of which is by using a combination of positive and negative selection

Some amino acids which have been added in order to label protein.

This orthologous set can then be mutated and screened through directed evolution to accept a different, even novel, amino acid. Mutations to the plasmid containing the pair can be introduced by error-prone PCR or through degenerate primers for the synthetase's active site.

Selection involves multiple rounds of a two-step process, where the plasmid is transferred into cells expressing chloramphenicol acetyl transferase with a premature amber codon.

In the presence of toxic chloramphenicol and the non-natural amino acid, the surviving cells will have overridden the amber codon using the orthogonal tRNA aminoacylated with either the standard amino acids or the non-natural one. To remove the former, the plasmid is inserted into cells with a barnase gene (toxic) with a premature amber codon but without the non-natural amino acid, removing all the orthogonal synthases which do not specifically recognize the non-natural amino acid.

In addition to the recoding of the tRNA to a different codon, they can be mutated to recognize a four base codon, allowing additional free coding options. The non natural amino acid, as a result, introduces diverse physicochemical and biological properties in order to be used as a tool to explore protein structure and function or to create novel or enhanced protein for practical purposes.

Diversity

The orthogonal pairs of synthase and tRNA which work for one organism may not work for another as the synthase may mis-aminoacylate endogenous tRNAs or the tRNA be mis-aminoacylated itself by an endogenous synthase. As a result the sets created to date different between organisms.

Orthogonal Sets in E. Coli

- tRNA-TyrRS pair from the archaeon *Methanococcus jannaschii*
- tRNA–LysRS pair from the archaeon *Pyrococcus horikoshii*
- tRNA–GluRS pair from *Methanosarcina mazei*
- leucyl-tRNA synthetase from *Methanobacterium thermoautotrophicum* and a mutant leucyl tRNA derived from *Halobacterium* sp.

Orthogonal Sets in Yeast

- tRNA-TyrRS pair from *Escherichia coli*
- tRNA–LeuRS pair from *Escherichia coli*
- tRNA from human and GlnRS from *Escherichia coli.*

Orthogonal Sets in Mammalian Cells

- tRNA-TyrRS pair from *Bacillus stearothermophilus*
- modified tRNA-TrpRS pair from *Bacillus subtilis* trp
- tRNA–LeuRS pair from *Escherichia coli.*

Protein Studies

With an expanded genetic code, the unnatural amino acid can be genetically directed to any chosen site in the protein of interest.

The high efficiency and fidelity allows a better control of the placement of the modification compared to modifying the protein post-translationally, which generally will target all amino acids of the same type, such as the thiol group of cysteine and the -amino group of lysine. Also, an expanded genetic code allows modifications to be carried out *in vivo*. The ability to site-specifically direct lab-synthesized chemical moieties into proteins allows many types of studies which would otherwise be extremely difficult.

- Probing Protein Structure and Function: by using amino acids with slightly different size such as o-Methyltyrosine or dansylalanine instead of tyrosine, and by inserting genetically-coded reporter moieties (color-changing and/or spin-active) into selected protein sites, chemical information about the protein's structure and function can be measured.
- Identifying and Regulating Protein Activity: by using photocaged aminoacids, protein function can be "switched" on or off by illuminating the organism.
- Changing the mode of action of a protein: one can start with the gene for a protein which binds a certain sequence of DNA, and, by inserting a chemically active amino acid into the binding site, convert it to a protein which cuts the DNA, rather than binding it.
- Improving immunogenicity and overcoming self-tolerance: by replacing strategically-chosen tyrosines with *p*-nitro phenylalanine, a tolerated self-protein can be made immunogenic.

An example of the possible application for this method is the biomedical where "chemical warheads" can be added to protein which target specific cellular components.

Since 2001, 40 non-natural amino acids have been added into protein by creating a unique codon (recoding) and a corresponding transfer-RNA:aminoacyl – tRNA-synthetase pair to encode it with diverse physicochemical and biological properties in order to be used as a tool to exploring protein structure and function or to create novel or enhanced proteins.

Theories on the Origin of the Genetic Code

Despite the minor variations that exist, the genetic code used by all known forms of life is nearly universal. However, there are a huge number of possible genetic codes. If amino acids are randomly associated with triplet codons, there will be 1.5 x 10 possible genetic codes. The question arises: why this code? How did it originate?

Phylogenetic analysis of transfer RNA suggests that tRNA molecules evolved before the present set of aminoacyl-tRNA synthetases.

Theoretically the genetic code could be completely random (a "frozen accident"), completely non-random (optimal) or a combination of random and nonrandom. There are sufficient data to refute the first possibility. For a start, a quick view on the table of the genetic code already shows a clustering of amino acid assignments.

Furthermore, amino acids that share the same biosynthetic pathway tend to have the same first base in their codons, and amino acids with similar physical properties tend to have similar codons. There are four themes running through the many theories that seek to explain the evolution of the genetic code (and hence the origin of these patterns):

- Chemical principles govern specific RNA interaction with amino acids. Aptamer experiments showed that some amino acids have a selective chemical affinity for the base triplets that code for them. Recent experiments show that of the 8 amino acids tested, 6 show some RNA triplet-amino acid association. This has been called the stereochemical code. The stereochemical code could have created an ancient core of assignments. The current complex translation mechanism involving tRNA and associated enzymes may be a later development, and that originally, protein sequences were directly templated on base sequences.
- Biosynthetic expansion. The standard modern genetic code grew from a simpler earlier code through a process of "biosynthetic expansion". Here the idea is that primordial life "discovered" new amino acids (e.g., as by-products of metabolism) and later back-incorporated some of these into the machinery of genetic coding. Although much circumstantial evidence has been found to suggest that fewer different amino acids were used in the past than today, precise and detailed hypotheses about exactly which amino acids entered the code in exactly what order have proved far more controversial.
- Natural selection has led to codon assignments of the genetic code that minimize the effects of mutations. A recent hypothesis suggests that the triplet code was derived from codes that used longer than triplet codons. Longer than triplet decoding has higher degree of codon redundancy and is more error resistant than the triplet decoding. This feature could allow accurate decoding in the absence of highly complex translational machinery such as the ribosome.
- The biocommunicative approach investigates nucleotide sequences as code, i.e. language-like text, which follows in parallel three (3) kinds of rules: combinatorial (syntactic), context-sensitive (pragmatic) and content-specific (semantic). Natural genome editing from this perspective is competent viral-driven generation and integration of meaningful nucleotide sequences into pre-existing genomic content arrangements and the ability to (re-)combine and (re-)regulate them according to context-dependent (i.e. adaptational) purposes of the host organism.
- Information-theoretic approaches treat the genetic code in terms of an error-prone information channel. The inherent noise in the channel poses the organism with a fundamental question: how to construct a genetic code that can withstand the impact of noise while accurately and efficiently translating information? These "rate-distortion" models suggest that the genetic code originated as a result of the interplay of

the three conflicting evolutionary forces: the needs for diverse amino-acids, for error-tolerance and for minimal cost of resources. The code emerges at a coding phase transition when the mapping of codons to amino-acids becomes nonrandom. The emergence of the code is governed by the topology defined by the most probable errors and is related to the map colouring problem.

- Genetic code and Codon populations: In other way, considering the global information level of Universal Genetic Code map, studying the whole Human genome codon populations show that the position of each codon within this table precisely dictates its population. So the Universal Genetic Code Table not only maps codons to amino acids, but serves perhaps as a global checksum matrix at the balance level of the whole genome: frequencies of the 64 codons in the whole human genome scale are a self-similar fractal expansion of the universal genetic code map. This global matrix-like function of Genetic Code table was published in September 2010 by the journal *Interdisciplinary Science* reporting that *Codon populations in single-stranded whole human genome DNA Are fractal and fine-tuned by the Golden Ratio 1.618.* In this paper, Jean-Claude Perez shows that Universal Genetic Code Table kernel governs not only the micro scale (codons i.e amino acids) but the macro scale as well. Particularly, the 6 folding steps of codon populations modeled by the binary divisions of the *Dragon fractal paper folding curve* extend the Chargaff's second rule from the domain of single TCAG nucleotides to the larger domain of codon triplets in the case of the whole Human genome.

Control of Gene Expression in Prokaryotes

Control of gene expression in prokaryotes involves control of "operons".

Operon consists of:

1. promoter - DNA sequence where RNA polymerase (transcription enzyme) binds.
2. operator - DNA sequence where an active repressor protein binds.
3. structural genes - DNA coding for protein (usually several that are functionally related, e.g. enzymes of a metabolic pathway).

Regulatory components (NOT part of the operon):

1. regulatory gene - DNA sequence coding for production of a repressor protein (often located away from operon it regulates).
2. repressor protein - in active form, it binds to operator to prevent transcription of structural genes.

Repressible Operon - ex. tryptophan operon : Repressible operons are usually in the "on" condition, i.e. their gene products are generally required most of the time for cell maintenance, etc. Repressor protein is made in an inactive form (it needs to bind to a corepressor to make it an active repressor which can then bind to operator).

Corepressor is usually an end-product of the metabolic pathway coded by the operon, e.g. tryptophan in our example here. Corepressor must accumulate to high levels before it can bind to inactive repressor and make it active.

Inducible Operon - ex. Lactose Operon

Inducible operons are usually in the "off" condition, i.e. their gene products are only occasionally (or never) needed by the cell.

Repressor protein is made in an active form (capable of binding to operator by itself).

Repressor protein can be inactivated by binding to an inducer (lactose in our example here) which then allows the operon to be activated.

Negative vs. Positive Controls

Above examples simply show operons being turned "on" or "off" at the operator, which represents a form of negative control.

Some operons, when turned "on", do not get transcribed at a rate sufficient to see a noticable effect because the RNA polymerase does not bind to promoter effectively.

These operons can be transcribed at a greatly accelerated rate by making the RNA polymerase binding much more efficient. This requires an activated "helper" for the RNA polymerase binding.

Gene Regulation in Eukaryotes

Developmental Biologists have long been interested in how a complex, multicellular organism is able to arise from a single fertilized egg (zygote), i.e. how cells arising from a common lineage (all arising from the zygote) are able to differentiate into the numerous, complex cell types that make up a multicellular organism.

Since the zygote contains ALL the genes necessary for the development of an organism, how is it that only a limited number of these genes are expressed in certain cells so that a multitude of cell types arise from the zygote, e.g. expression of genes for nerve cells, bone cells, skin cells, etc.

At one time, it was generally believed that progressive determination was caused by groups of cells actually losing most of the zygotes original set of genes until a particular cell type contained only genes for forming that cell type, e.g. a cell that eventually becomes a nerve cell has gradually lost all of the genes of the original zygote except those for nerve cell formation.

Today we know that this is generally not the case, although there are exception.

Mitosis faithfully reproduces the full chromosome complement (and all of their genes) to be equally distributed to the new daughter cells, i.e. the daughter cells retain the total genetic potential of the original zygote, "totipotent cells".

Much experimental evidence now supports this concept of the genetic equivalence of nuclei in a multicellular organism.

Nuclear transplantation experiments performed by Robert Briggs and Thomas King, and John Gurdon showed that cells of early embryos retain full genetic potential of zygote. F.C. Steward was the first to show that individual cells taken from a mature carrot plant could grow into new carrot plants (clones). More recently the cloning of "Dolly" the sheep, which was not all that important scientifically except that it showed that it could be done in mammals (humans?).

Regeneration, in both plants and animals, have long supported the concept of totipotent cells or the genetic equivalence of somatic cell nuclei.

Therefore, the fact that a differentiating cell is using only a limited number of the total available genes, the problem becomes a differential activation/inactivation of available genes.

For example, cells that eventually become nerve cells do so by activating only the genes necessary for nerve cell formation and likewise inactivating all other genes.

The ways in which cells control gene expression has become a major focus of biological research in the last half of the 20th Century and with the more recent genetic engineering techniques will continue to be a major focus well into this century.

Most of what we first learned about differential gene activation/inactivation came from work on bacteria, e.g. *Escherichia coli*.

Although these cells do not undergo the complex developmental phenomena associated with multicellular organism, they do require activation and inactivation of genes to maintain their existence, and their relative simplicity lent them very well to early studies of gene control.

Many of the models for gene control in prokaryotes have been used as model systems to initially investigate eukaryotic gene control.

However, as one might suspect, the controls used to regulate euakyotic genes appear to be much more complex and must take into account certain aspects of the eukaryote genome that are non-existent in prokaryotes.

1. Very large number of genes present (30,000+) human genes; although only about 3-5% is being used at any one time; recall that *E. coli* has only about 2500 genes.
2. Most eukaryote DNA does not code for RNA or protein (does not function as "genes").
3. Presence of chromatin and the structural complexity of the chromosome.
4. Compartmentalization of most DNA into a nucleus.
5. Complexity in variety of genes being activated at any one time (recall functionally linked genes into operons in bacteria), etc.

One way of looking at "gene expression" is to consider that a gene is expressed (or "activated") only when the product that the gene codes for is produced and functional, e.g. an "active" enzyme.

Considering this very broad description for "gene activation", the eukaryotic cell possesses numerous places where control could take place. Some of these are, at present, only possibilities while others have good experimental support for their existence.

Potential Levels of Eukaryotic Gene Control

Pretranscriptional Control

The physical form of eukaryotic chromatin may be able to block the availability of genes for transcription by forming chromatin structures that "hide" the DNA from trancription factors and enzymes.

"DNA packing" involved basic proteins called HISTONES (H1, H2A, H2B, H3, H4) and various other chromosomal proteins.

An extreme form of DNA packing occurs when chromosomes condense during Prophase of cell division. Very little, if any, gene activity occurs while mitotic/meiotic chromosomes are formed.

Heterochromatin is chromatin in a highly compacted form that is very low or completely devoid of transcriptional activity. It can occur in chromatin found in the Interphase cell.

The tight coiling might be due to high methylation of the DNA in these areas.

Regardless of the cause of heterochromatin, experimental evidence has shown very little transcriptional activity associciated with these areas.

DNA methylation, in general, is a means of inactivating genes over extended time. An extreme example of gene inactivation by heterochromatin formation is the Barr body typically found in human females. In this example, an entire chromosome's genes are prevented from being transcribed (activated).

A chromosome may have heterochromatin areas interspersed with less compacted areas called euchromatin, which are thought to be areas where the DNA is more available for transcription.

Even in euchromatin, the DNA is tightly wrapped around nucleosomes formed by histones. To make this DNA available for transcription, the histones of the nucleosome appear to undergo acetylation (addition of acetyl groups) which causes a "relaxation" of the DNA/nucleosome association.

This relaxing of the DNA/nucleosome assoication makes the DNA more available for transciption.

Transcriptional Control

Regulatory proteins have also been shown to exist in eukaryotes and may be either activators or repressors of gene activity.

Since the majority of the eukaryotic genes are inactive at any one time, it is believed that most regulators are activators.

Enhancers have also been shown to exist in eukaryotes. These are small stretches of DNA that have the ability to enhance transcriptional activity after binding activator proteins and then interacting at the promoter region to enhance RNA polymerase binding and activity.

Recall that transcription factors are also needed for RNA polymerase binding and activity.

Enhancers are found some distance away from the promoter and may be "upstream" or "downstream" from the promoter.

Posttranscriptional Control

The initial transcript (RNA) in eukaryotes is not functional but must undergo modification to become functional.

rRNA, tRNA, and mRNA must all undergo varying degrees of posttranscriptional modification before they can become functional.

In general, these modifications may include:

1. the cutting out of functional pieces of RNA from a much larger nonfunctional transcript.
2. some of the above pieces may be functional but some require the bonding together of excised pieces to form a single functional piece (exons/introns).

3. mRNA requires the most extensive posttranscriptional modifications, which in addition to the above, require chemical modifications at both ends of the molecule. These modifications may help the transport out of the nucleus, protect the molecule from nuclease attack in the cytoplasm, or somehow be required for initiation of translation.

mRNA is generally consider a "short-lived" molecule in both prokaryotes and eukaryotes, and was once used as a prime criterion for proving that a RNA was in fact a mRNA.

However, exceptions to this generality have been shown in eukaryotes, particularly in certain highly specialized cells (e.g. maturing vertebrate erythrocytes) and developing embryos.

In the case of the maturing vertebrate red blood cell (called erythropoiesis), the erythroblast is a nucleated cell that gradually shuts down most of it's genes except for mRNA for globin production. Huge amounts of globin mRNA accumulate and eventually the cell becomes anucleate, however, the globin mRNA remains for a considerable time afterwards while the cell makes and accumulates very large amounts of hemoglobin.

Unfertilized eggs have been shown to manufacture and accumulate large amounts of mRNA coding for proteins necessary for the rapid, early development of the embryo; these mRNAs are stored in an inactive form in the egg. Activation of the egg, by sperm or artificial means, activates these mRNAs for embryo protein production.

Translational Control

The observation that eggs may contain "long-lived" mRNA but that this RNA is not translated until egg activation has been shown to be a block at translation. A certain initiation factor(s) necessary for translation rapidly becomes available after fertilization (activation).

Also, the amount of heme remaining in maturing red blood cells helps determine how much more globin will be made (translation of globin mRNA).

The poliovirus is thought to select the translation of its "uncapped" mRNAs by somehow destroying the host cell's cap recognition system on the ribosome complex.

Posttranslational Control

Even after the protein is first made, it is not necessarily functional.

Insulin is initially made as a single protein (inactive proinsulin) which must fold back upon itself, form three disulfide bonds, and finally have a small stretch of amino acids removed by a peptidase before it becomes active insulin consisting of two polypeptides held together by two of the disulfide bonds.

Many of the digestive enzymes of the small intestine are synthesized as zymogens (inactive forms) that must undergo enzymatic cleavage to become active, e.g. trypsinogen/trypsin, chymotrypsinogen/chymotrypsin, pepsinogen/pepsin.

Some proteins (malformed, no longer needed, etc.) are "marked" for degradation by proteosomes.

Eukaryotic Gene Organization and its Evolution

Unlike the prokaryotes, eukaryotes do not possesss an "operon type" arrangement of functionally related genes. (recall prokaryote operon).

Instead, it has been shown that genes coding for functionally related products, e.g. a series of enzymes in a metabolic pathway, are scattered throughout the genome and each controlled by a separate promoter, etc.

This gene arrangement may at first appear disorganized but the reality is that it allows much greater variety in the use of these "scattered" genes. Genes can be activated in numerous combinations.

A multigene family is a collection of genes within a genome that are similar or identical in sequence.

The 28S, 18S and 5.8S rRNA genes occur in eukaryotes as tandem arrays of several hundred to several thousand located in association with a nucleosome found inside the nucleus.

The family of globin genes are sequence related but not identical and function in making different hemoglobins at different developmental stages.

Both types of multigene families most likely arose by gene duplication and mutation.

Molecular Biology of Cancer

Cancer is caused by inapproriate gene activity associated with the cell cycle.

Some chararcteristics of malignant cancers (solid types):

1. Loss of cell cycle control
2. Cell surface alterations
3. Secretion of distinct products, e.g. proteolytic enzymes, embryonic antigens, angiogenin, etc.
4. Morphological alterations, enlarged nuclei, aberrant nucleoli, numerous microvilli
5. Highly vascularized
6. Metastasis (travel to distant tissues and establish secondary or "metastatic" cancers).

The genetic basis of cancer was first discovered through the work with retroviruses that were shown to cause cancer in monkeys, chickens, mice, etc.

The viral genes were identified that could be directly linked to the cause of cancer in experimental animals.

These genes were called oncogenes ("cancer genes").

Upon closer examination, researchers found that in fact the oncogenes were not unique but were in fact mutated forms of normal genes (proto-oncogene) found in eukaryotic organisms; including humans.

These normal genes, for the most part, are involved with some aspect of regulating the cell cycle.

Three ways for conversion of a proto-oncogene into an oncogene:

1. translocation or transposition to new promoter
2. gene amplification
3. point mutation within gene.

In addition to proto-oncogenes that can become oncogenes, there are normal, tumour-suppressor genes that are mutated during the formation of cancers.

Tumour-suppressor genes produce products that normally block the cell cycle; which would normally prevent cancers.

The cancerous state results from alteration of several genes, both proto-oncogenes and tumour suppressor genes.

2

Chromosome Mutations

It is natures intention that the exact genetic information from both parents will be seen in the offspring's DNA in the the critical stages of fertilisation. However, it is possible for this genetic information to mutate, which in most cases, can result in fatal or negative consequencies in the outcome of the new ogranism.

Non-Disjunction and Down's Syndrome

One well known example of mutation is non-disjunction. Non-disjunction is when the spindle fibres fail to seperate during meiosis, resulting in gametes with one extra chromosome and other gametes lacking a chromosome.

If this non-disjunction occurs in chromosome 21 of a human egg cell, a condition called Down's syndrome occurs. This is because their cells possess 47 chromosomes as opposed to the normal chromosome compliment in humans of 46.

The fundamental structure of a chromosome is subject to mutation, which will most likely occur during crossing over at meiosis. There are a number of ways in which the chromosome structure can change, as indicated below, which will detrimentally change the genotype and phenotype of the organism.

However, if the chromosome mutation effects an essential part of DNA, it is possible that the mutation will abort the offspring before it has the chance of being born.

The following indicates types of chromosome mutation where whole genes are moved:

Deletion of a Gene

As the name implies, genes of a chromosome are permanently lost as they become unattached to the centromere and are lost forever

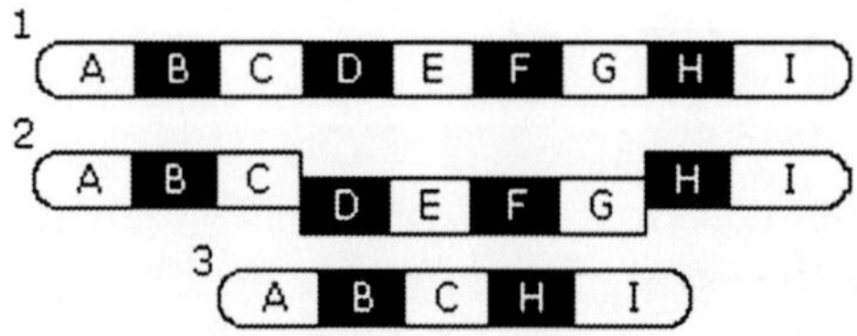

- Normal chromosome before mutation
- Genes not attached to centromere become loose and lost forever
- New chromosome lacks certain genes which may prove fatal depending on how important these genes are

Duplication of Genes

In this mutation, the mutants genes are displayed twice on the same chromosome due to duplication of these genes. This can prove to be an advantageous mutation as no genetic information is lost or altered and new genes are gained

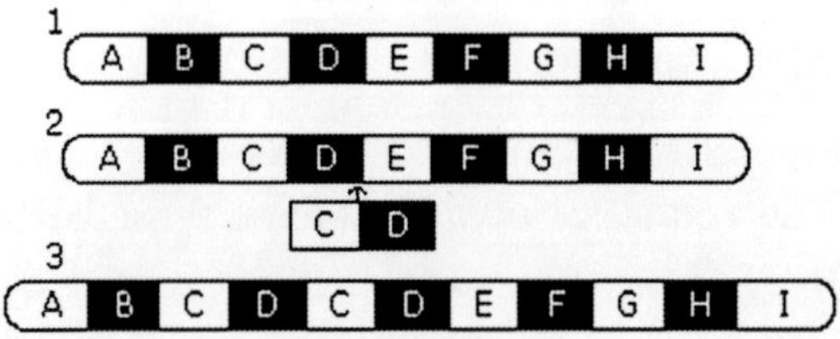

- Normal chromosome before mutation
- Genes from the homologous chromosome are copied and inserted into the genetic sequence
- New chromosome possesses all its initial genes plus a duplicated one, which is usually harmless

Inversion of Genes

This is where the order of a particular order of genes are reversed as seen below

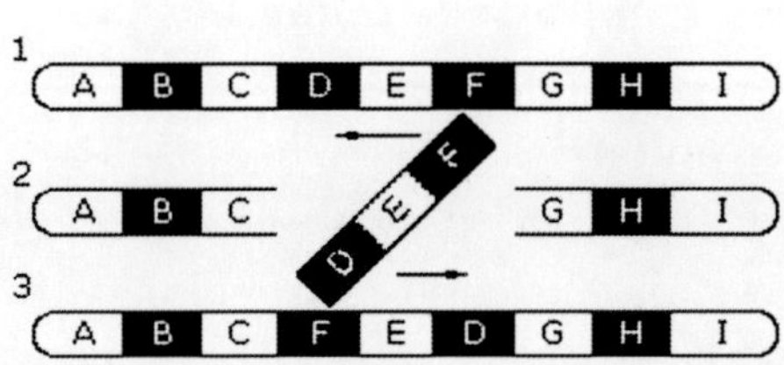

- Normal chromosome un-altered

- The connection between genes break and the sequence of these genes are reversed
- The new sequence may not be viable to produce an organism, depending on which genes are reversed. Advantageous characteristics from this mutation are also possible

Translocation of Genes

This is where information from one of two homologous chromosomes breaks and binds to the other. Usually this sort of mutation is lethal

- An un-altered pair of homologous chromosomes
- Translocation of genes has resulted in some genes from one of the chromosomes attaching to the opposing chromosome

Alteration of a DNA Sequence

The previous examples of mutation have investigated changes at the chromosome level. The sequence of nucleotides on a DNA sequence are also susceptible to mutation.

Deletion: Here, certain nucleotides are deleted, which affects the coding of proteins that use this DNA sequence. If for example, a gene coded for alanine, with a genetic sequence of C-G-G, and the cytosine nucleotide was deleted, then the alanine amino acid would not be able to be created, and any other amino acids that are supposed to be coded from this DNA sequence will also be unable to be produced because each successive nucleotide after the deleted nucleotide will be out of place.

Insertion: Similar to the effects of deletion, where a nucleotide is inserted into a genetic sequence and therefore alters the chain thereafter. This alteration of a nucleotide sequence is known as frameshift

Inversion: Where a particular nucleotide sequence is reversed, and is not as serious as the above mutations. This is because the nucleotides that have been reversed in order only affect a small portion of the sequence at large

Substitution: A certain nucleotide is replaced with another, which will affect any amino acid to be synthesised from this sequence due to this change. If the gene is essential, i.e. for the coding of haemoglobin then the effects are serious, and organisms in this instance suffer from a condition called sickle cell anaemia.

All of the genetic mutations looked at through the last 2 pages more or less have a negative impact and are undesired, however, in some cases they can prove advantageous.

Genetic mutations increase genetic diversity and therefore have an important part to play. They are also the reason many people inherit diseases.

Mutation Frequency and Polyploidy

The previous two pages have investigated mutations, and this page continues with more information related to genetic mutations.

Polyploidy

Humans are diploid creatures, meaning for every chromosome in our body, there is another one to match it. Read the following

- Haploid creatures have one of each chromosome
- Diploid creatures have two of each chromosome
- Triploid creatures have three of each chromosome
- Polyploid creatures have three or more of each chromosome
- They can be represented by n where n equals haploid, 2n equals diploid and so on.

It is possible for a species, particularly plant species, to produce offspring that contains more chromosomes than its parent. This can be a result of non-disjunction, where normally a diploid parent would produce diploid offspring, but in the case of non-disjunction in one of the parents, produces a polyploid.

In the case of triploids, although the creation of particular triploids in species is possible, they cannot reproduce themselves because of the inability to pair homologous chromosomes at meiosis, therefore preventing the formation of gametes.

Polyploidy is responsible for the creation of thousands of species in today's planet, and will continue to do so. It is also responsible for increasing genetic diversity and producing species showing an increase in size, vigour and an increased resistance to disease.

Mutation Frequency

This page and the previous two have investigated the different ways that mutations arise, and the following elaborates on the ways in which mutations are instigated.Barring all external factors, mutations occur very rarely, and are rarely expressed because many forms of mutation are expressed by a recessive allele. However there are many mutagenic agents that artificially increase the rate of mutations in an organism. The following are some factors that increase genetic mutations in organisms

- Members of species in a particular geographic area or ethnic origin are more susceptible to mutations
- High dosages of X-Rays or ultraviolet light can increase the likeliness of a mutation
- Radioactive substances increase the rate of mutations exponentially

As mentioned previously, genetic mutations are a source of new variation in a species because it physically alters the sequence of nucleotides in a given sequence, therefore altering the genome in a unique way.

Theory of Natural Selection

In the 19th century, a man called Charles Darwin, a biologist from England, set off on the ship HMS Beagle to investigate species of the island. After spending time on the islands, he soon developed a theory that would contradict the creation of man and imply that all species derived from common ancestors through a process called natural selection. Natural selection is considered to be the biggest factor resulting in the diversity of species and their genomes. The principles of Darwin's work and his theory are stated below.

- One of the prime motives for all species is to reproduce and survive, passing on the genetic information of the species from generation to generation. When species do this they tend to produce more offspring than the environment can support.
- The lack of resources to nourish these individuals places pressure on the size of the species population, and the lack of resources means increased competition and as a consequence, some organisms will not survive.
- The organisms who die as a consequence of this competition were not totally random, Darwin found that those organisms more suited to their environment were more likely to survive.
- This resulted in the well known phrase survival of the fittest, where the organisms most suited to their environment had more chance of survival if the species falls upon hard times. (This phrase if often associated with Darwin, though on closer inspection Herbert Spencer puts the phrase in a more accurate historical context.)
- Those organisms who are better suited to their environment exhibit desirable characteristics, which is a consequence of their genome being more suitable to begin with.

This 'weeding out' of less suited organisms and the reward of survival to those better suited led Darwin to deduce that organisms had evolved over time, where the most desirable characteristics of a species are favoured and those organisms who exhibit them survive to pass their genes on.

As a consequence of this, a changing environment would mean different characteristics would be favourable in a changing environment. Darwin believed that organisms had 'evolved' to suit their environments, and occupy an ecological niche where they would be best suited to their environment and therefore have the best chance of survival. As the above indicates, those alleles of a species that are favoured in the environment will become more frequent in the genomes of the species, due to the organisms higher likeliness of surviving as part of the species at large

Recombinant DNA

Recombinant DNA (rDNA) molecules are DNA sequences that result from the use of laboratory methods (molecular cloning) to bring together genetic material from multiple sources, creating sequences that would not otherwise be found in biological organisms. Recombinant DNA is possible because DNA molecules from all organisms share the same chemical structure; they differ only in the sequence of nucleotides within that identical overall structure. Consequently, when DNA from a foreign source is linked to host sequences that can drive DNA replication and then introduced into a host organism, the foreign DNA is replicated along with the host DNA. Recombinant DNA molecules are sometimes called chimeric DNA, because they are usually made of material from two different species, like the mythological chimera.

The DNA sequences used in the construction of recombinant DNA molecules can originate from any species. For example, plant DNA may be joined to bacterial DNA, or human DNA may be joined with fungal DNA. In addition, DNA sequences that do not occur anywhere in nature may be created by the chemical synthesis of DNA, and incorporated into recombinant molecules.

Using recombinant DNA technology and synthetic DNA, literally any DNA sequence may be created and introduced into any of a very wide range of living organisms.

Proteins that result from the expression of recombinant DNA within living cells are termed recombinant proteins. When recombinant DNA encoding a protein is introduced into a host organism, the recombinant protein will not necessarily be produced . Expression of foreign proteins requires the use of specialized expression vectors and often necessitates significant restructuring of the foreign coding sequence .

It is important to note that recombinant DNA differs from genetic recombination in that the former results from artificial methods in the test tube, while the latter is a normal biological process that results in the remixing of existing DNA sequences in essentially all organisms.

Creating Recombinant DNA

Construction of recombinant DNA, in which a foreign DNA fragment is inserted into a plasmid vector. In this example, the gene indicated by the white color is inactivated upon insertion of the foreign DNA fragment.

Molecular cloning is the laboratory process used to create recombinant DNA. It is one of two widely-used methods (along with polymerase chain reaction, abbr. PCR) used to direct the replication of any specific DNA sequence chosen by the experimentalist. The fundamental difference between the two methods is that molecular cloning involves replication of the DNA within a living cell, while PCR replicates DNA in the test tube, free of living cells.

Formation of recombinant DNA requires a cloning vector, a DNA molecule that will replicate within a living cell. Vectors are generally derived from plasmids or viruses, and represent relatively small segments of DNA that contain necessary genetic signals for replication, as well as additional elements for convenience in inserting foreign DNA, identifying cells that contain recombinant DNA, and, where appropriate, expressing the foreign DNA.

The choice of vector for molecular cloning depends on the choice of host organism, the size of the DNA to be cloned, and whether and how the foreign DNA is to be expressed. The DNA segments can be combined by using a variety of methods, such as restriction enzyme/ligase cloning or Gibson assembly. In standard cloning protocols, the cloning of any DNA fragment essentially involves seven steps: (1) Choice of host organism and cloning vector, (2) Preparation of vector DNA, (3) Preparation of DNA to be cloned, (4) Creation of recombinant DNA, (5) Introduction of recombinant DNA into the host organism, (6) Selection of organisms containing recombinant DNA, (7) Screening for clones with desired DNA inserts and biological properties. These steps are described in some detail in a related article (molecular cloning).

Expression of Recombinant DNA

Following transplantation into the host organism, the foreign DNA contained within the recombinant DNA construct may or may not be expressed. That is, the DNA may simply be replicated without expression, or it may be transcribed and translated so that a recombinant protein is produced. Generally speaking, expression of a foreign gene requires restructuring the gene to include sequences that are required for producing a mRNA molecule that can be used by the host's translational apparatus (e.g. promoter, translational initiation signal, and transcriptional terminator). Specific changes to the host organism may be made to improve expression of the ectopic gene. In addition, changes may be needed to the coding sequences as well, to optimize translation, make the protein soluble, direct the recombinant protein to the proper cellular or extracellular location, and stabilize the protein from degradation.

Properties of Organisms Containing Recombinant DNA

In most cases, organisms containing recombinant DNA have apparently normal phenotypes. That is, their appearance, behavior and metabolism are usually unchanged, and the only way to demonstrate the presence of recombinant sequences is to examine the DNA itself, typically using a polymerase chain reaction (PCR) test.

If the rDNA sequences encode a gene that is expressed, then the presence of RNA and/or protein products of the recombinant gene can be detected, typically using RT-PCR or western hybridization methods. Gross phenotypic changes are not the norm, unless the recombinant gene has been chosen and modified so as to generate biological activity in the host organism.

Additional phenotypes that are encountered include toxicity to the host organism induced by the recombinant gene product, especially if it is over-expressed or expressed within inappropriate cells or tissues.

In some cases, recombinant DNA can have deleterious effects even if it is not expressed. One mechanism by which this happens is insertional inactivation, in which the rDNA becomes inserted into a host cell's gene. In some cases, researchers use this phenomenon to "knock out" genes in order to determine their biological function and importance. Another mechanism by which rDNA insertion into chromosomal DNA can affect gene expression is by inappropriate activation of previously unexpressed host cell genes. This can happen, for example, when a recombinant DNA fragment containing an active promoter becomes located next to a previously silent host cell gene, or when a host cell gene that functions to restrain gene expression undergoes insertional inactivation by recombinant DNA.

Recombinant DNA Technology

All gene manipulation is based on microbial genetics—ways of doing in the test tube what bacteria and viruses do naturally. Several of this week's gene manipulations are exemplified by the article on overexpressed angiopoietin in a transgenic mouse.

Gel Electrophoresis

This technique separates molecules on the basis of their size.

- Cast slab of gel material, usually agarose or polyacrylamide. The gel is a matrix of polymers forming sub-microscopic pores.

- The size of the pores can be controlled by varying the chemical composition of the gel.

The gel is set up for electrophoresis in a tank holding pH buffer. Electrodes apply an electric field:

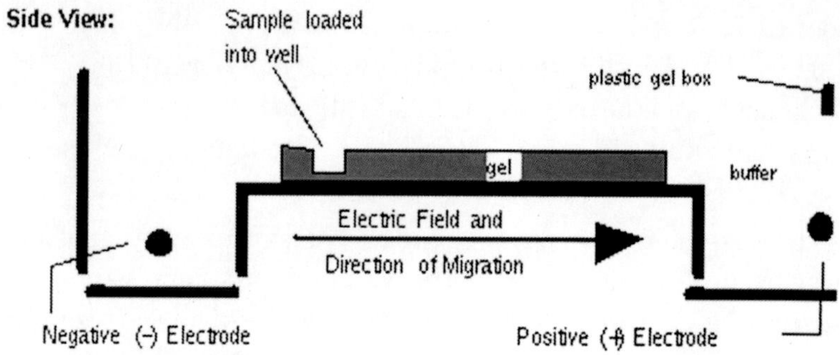

The molecules to separate (DNA RNA) carry a net negative charge (why?) so they move along the electric field toward the positive cathode. (To separate proteins, a detergent would be included which coats the protein with negative charge.) The larger molecules are held up as they try to pass through the pores of the gel, while the smaller molecules are impeded less and move faster.

This results in separation by size, with the larger molecules nearer the well and the smaller molecules farther away.

Note that this separates on the basis of size (volume in solution), which is not necessarily molecular weight. For example:

- Two DNA molecules of the same molecular weight will run differently if one is supercoiled, because the supercoils constrain the shape to be smaller.
- Two RNA molecules of the same molecular weight will run differently if one has much intramolecular base pairing, making it "smaller."

Aside from the above exceptions, the distance migrated is roughly proportional to the log of the inverse of the molecular weight (the log of 1/MW). Gels are normally depicted as running vertically, with the wells at the top and the direction of migration downwards. This leaves the large molecules at the top and the smaller molecules at the bottom. Molecular weights are measured with different units for DNA, RNA, and protein:

- DNA: Molecular weight is measured in base-pairs, or bp, and commonly in kilobase-pairs (1000bp), or kbp.
- RNA: Molecular weight is measured in nucleotides, or nt, and commonly in kilonucleotides (1000nt), or knt. [Sometimes, bases, or b and kb are used.

- Protein: Molecular weight is measured in Daltons (grams per mole), or Da, and commonly in kiloDaltons (1000Da), or kDa.

Molecular weight standards run in one well of the gel are used to calibrate the molecular weights of sample molecules. Below is a gel stained with a dye: a colored molecule which binds to a specific class of macromolecules in a sequence-independent manner (probes bind in a sequence-dependent manner).

Sample 1 contains only one size class of macromolecule - it could be a plasmid, a pure mRNA transcript, or a purified protein. In this case, you would not have to use a probe to detect the molecule of interest since there is only one type of molecule present. Blotting is usually necessary for samples that are not complex mixtures. By interpolation, its molecular weight is roughly 3.

Sample 2 is what a sample of total DNA cut with a restriction enzyme, total cellular RNA, or total cellular protein would look like in a gel stained with a sequence-independent stain.

There are so many bands that it is impossible to find the one we are interested in. Without a probe (which acts like a sequence-dependent stain) we cannot get very much information from a sample like this.

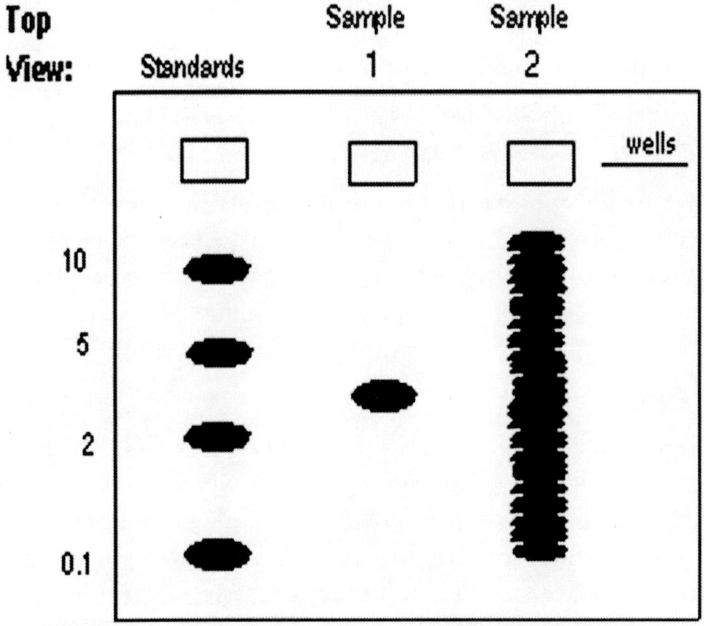

Different stains are used for different classes of macromolecules. DNA and RNA are generally stained with ethidium bromide (EtBr), an intercalating agent. The DNA-EtBr complex fluoresces under UV light. Protein is stained with Coomassie Blue or Silver Stain.

Cloning Genes

In nature, DNA molecules recombine for various functions — even DNA between different species. But twenty years ago, despite the work of Barbara McClintock and others, the extent of this recombination was not appreciated. DNA was still thought to be the "master molecule," not to be violated by "unnatural" manipulation. When scientists began to manipulate DNA in the test tube, many scientists feared that disastrous monsters would result, with unspecified dangers to people.

In 1977 scientists at the Asilomar Conference proposed sweeping regulation on so-called "recombinant DNA," technologies which recombine DNA from different species in the test tube.

Since then, the dangers have appeared to be little more than those of "natural" genetic mixing. But we remain concerned about issues such as:

- Engineering food crops to resist pesticides. The pesticide resistance genes can escape into natural populations of weeds.
- Engineering a human symbiont microbe, such as E. coli, to produce a deadly toxin such as botulin. In theory this could be done, although it's not clear where such an organism would live, or how well it could "compete" with natural flora.
- Societal dilemmas of human cloning. How far shall we use reproductive technology to shape future humans?

Techniques of Recombinant DNA

How do we manipulate these natural processes for biotechnology; for instance, to make a bacterium that produces large quantities of insulin? One approach would be to cut the appropriate gene from human DNA and paste, or splice, it into a vector such as a plasmid or phage DNA. Our "scissors" are the class of enzymes called restriction endonucleases

Restriction Endonucleases

An "endonuclease" is an enzyme that cuts duplex DNA in the middle, not at an end (for exonuclease). Different species of bacteria have evolved different restriction endonucleases, each to cut foreign DNA that gets into their cells by mistake. To be cut, the DNA has to lack their own pattern of protective methylation. There are well over a hundred restriction enzymes, each cutting in a very precise way a specific base sequence of the DNA molecule.

A restriction endonuclease cuts DNA only at a specific site, usually containing 4-6 base pairs. The enzyme has to cut the DNA backbone twice, recognizing the same type of site; therefore, the site "reads" the same way backwards as forwards—a palindrome.

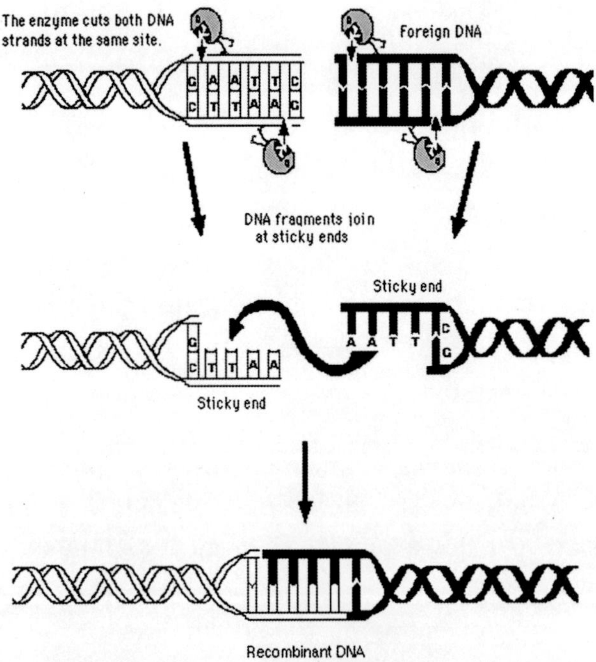

Figure: *Restriction Enzyme Action of EcoRI*

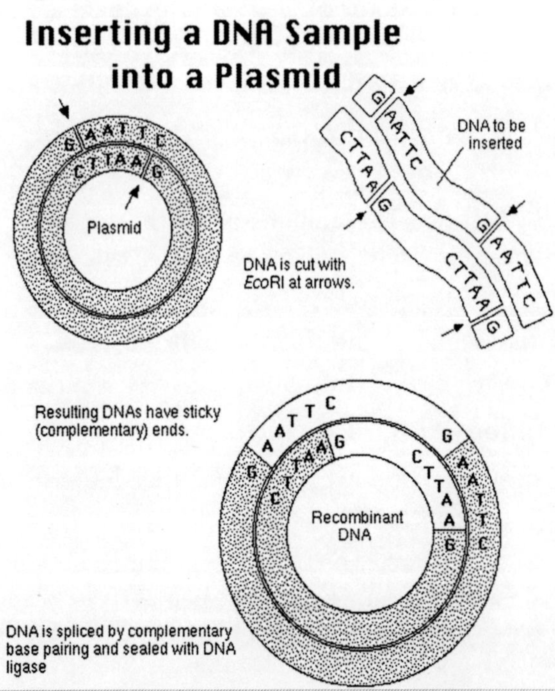

This "sticky ends" from two different DNA molecules can hybridize together; then the nicks are sealed using ligase. (Where does ligase come from? What is its natural function?) The result is recombinant DNA. When this recombinant vector is inserted into E. coli, the cell will be able to process the instructions to assemble the amino acids for insulin production.

More importantly, the new instructions are passed along to the next generation of E. coli cells in the process known as gene cloning.

Restriction Site Analysis

How can we use restriction sites to analyze the plasmid products of ligation, and tell whether we in fact have ligated the correct molecule:

Problem: Suggest several "incorrect" ways the plasmid could recombine. Use the PLASMID Program.

You MUST practice restriction analysis with this program; it will be on the quiz and/or the test. How do we get the recombinant molecule into a bacterial cell? Usually by transformation (for a plasmid) or by in vitro packaging into a phage head coat (for a phage vector such as lambda phage).

How would we actually locate the appropriately cloned gene? There are many different ways, depending on the specific case.

Here is one example, in which a partial sequence of the protein enables us to reverse the code and determine an approximate DNA sequence to use for a radiolabeled probe. The DNA probe will hybridize to clones containing the correct DNA, even if it is just one piece cut out of an entire genome.

Griffiths et al, W. H. Freeman & Co., current edition

The radioactive probe is made by determining a short segment of the protein sequence, then "back translating" to the possible DNA sequences. Short DNA sequences are synthesized to match the protein sequence.

Then these DNA oligomers (known as "oligos") are radiolabeled, and applied to the blotted clones. They should hybridize only to clones containing sequence encoding the desired protein.

Reverse Transcription: cDNA Cloning

Suppose we need to clone a gene containing lots of introns. What will happen when the bacterium tries to express it?

To overcome this problem, we can start with mRNA isolated from tissues that produce the desired protein. We then use reverse transcriptase enzyme (produced by a retrovirus related to HIV) to reverse transcribe the mRNA into a DNA molecule that now is free of introns.

Now we can ligate "sticky ends" onto the cDNA and recombine it into a phage or plasmid vector.

Problem: Know the differences between genomic cloning and cDNA cloning. Explain the relative ADVANTAGES and DISADVANTAGES of each technique—depending on the aim of your research.

Gene Therapy

Transgenic Plants

In PCR, a heat-stable DNA polymerase is used, most commonly Taq Polymerase from the thermophilic microbe Thermus aquaticus. Thomas Brock discovered T. aquaticus from a hot spring at Yellowstone National Park.

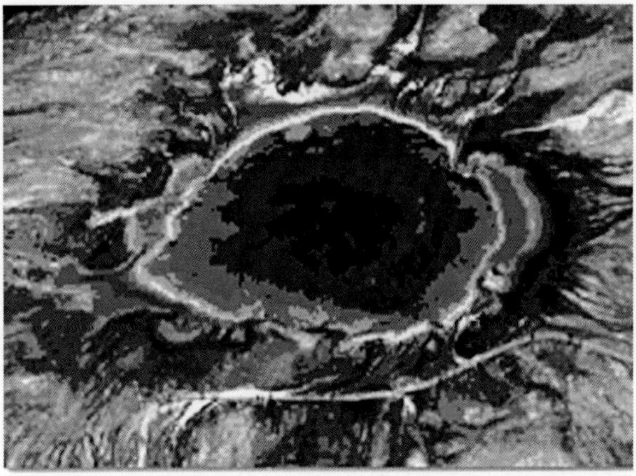

More recently, an even more heat-resistant polymerase has been developed from a hyperthermophilic microbe growing at 110 degrees C in hydrothermal vent ecosystems in the deep ocean; it's called "Vent Polymerase."

The Taq Polymerase is put with the DNA to be amplified, plus all four NTPs, plus two primers facing each other, about 200 - 6000 kb apart. (Why do we need primers?) The primers are selected based on the DNA region you want to amplify.

The tube is placed in a thermal cycler. DNA gets synthesized from each primer, for about 2 minutes.

Then the temperature is raised to 95 degrees C — enough to denature (split apart) the DNA base pairs. But the Taq Polymerase remains intact, because it comes from an organism that evolved to grow at this temperature. Now the temperature is decreased again, and primers again can hybridize to the DNA—both the old AND the newly synthesized strands.

Again, Taq Polymerase extends new DNA strands. Again, the temperature is raised. After repeated cycles, the amount of DNA sequence between the two primers increases exponentially. First 2 strands, then 4, 8, 16, up to about a million. Thus, in a couple of hours, you can get million-fold amplification of a DNA sequence.

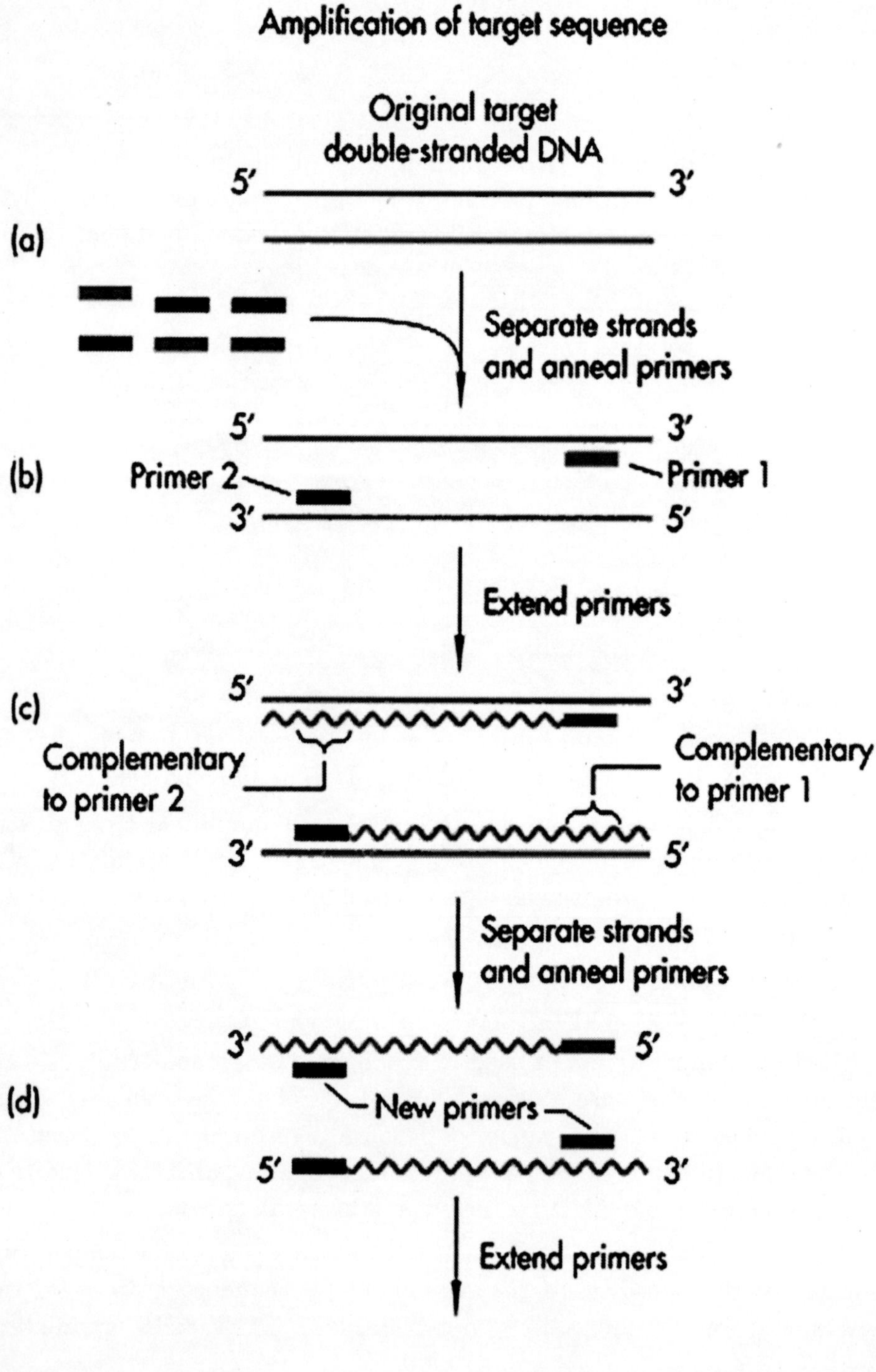

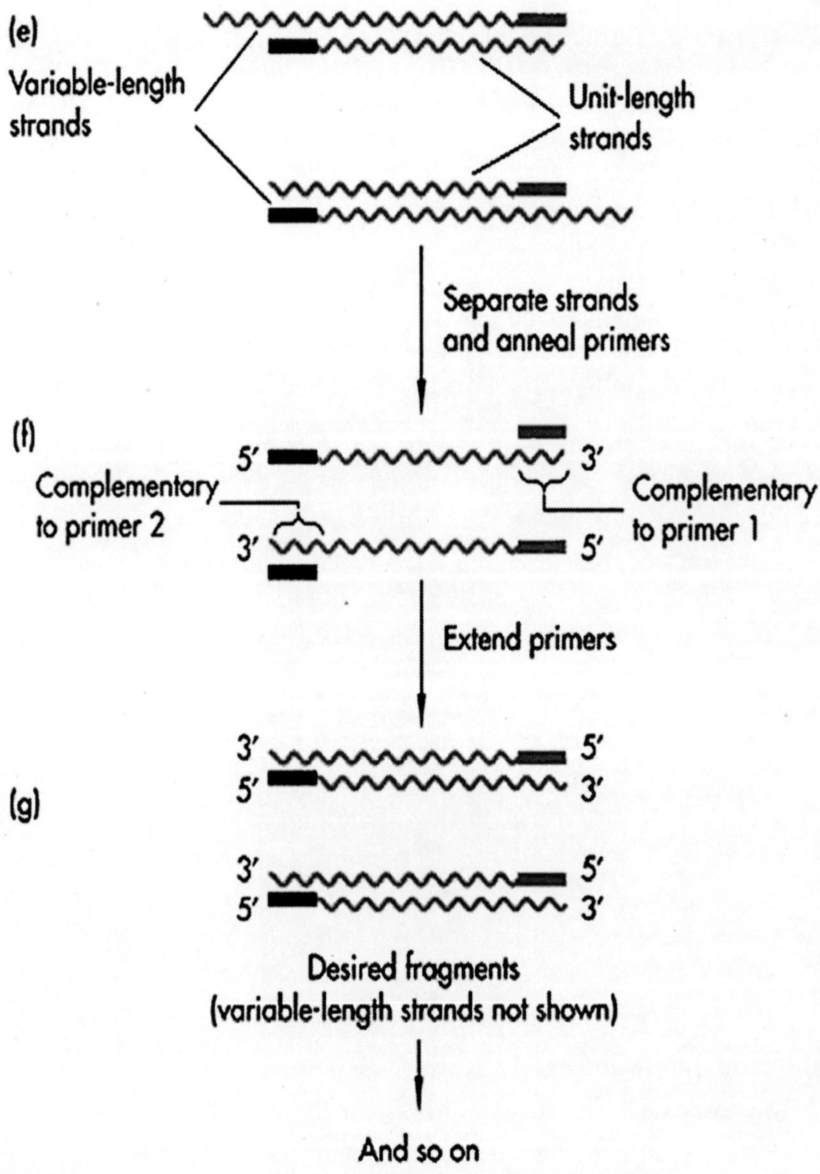

Griffiths et al, W. H. Freeman & Co., current edition

Applications of PCR

PCR has replaced cloning for many purposes, particularly the sequencing of DNA. It is faster and requires no vectors, which can mutate as they reproduce. It can be used forensically, to amplify tiny amounts of DNA from criminal evidence; or clinically, to detect DNA sequences linked to inherited disorders.

PCR Experiment

Microbiology Lab: Students characterized the microbial species of various environments in Knox County. Microbial colonies were placed directly into PCR reactions containing primers for amplification of ribosomal RNA genes (rDNA). Lane 4 contains an amplified band of the predicted size, 1000 bp. The top of each well contains genomic DNA. The smears at the bottom contain the PCR primers. The DNA will be sequenced and matched through GenBank to determine the microbial genus.

Thanks to Dan Nickerson '00 and Adam Marks '01.

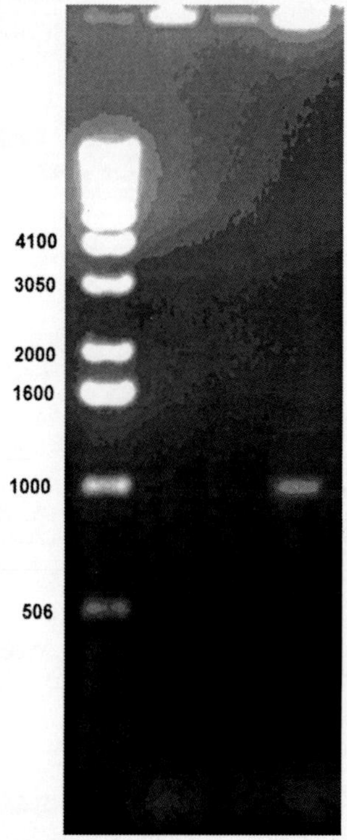

The main limitations of PCR are:

- Only relatively short sequences can be amplified reliably. Anything more than 10,000 base pairs is unlikely to be amplified.
- You need to know the right primer sequences to use, at both ends of the sequence you want to amplify. If two related genes have the same end sequences, you might amplify the wrong gene.
- You only obtain a DNA fragment. To see this DNA at work inside a living organism, some type of cloning has to be done.
- PCR Technology - discussion of preparation of the sample, the master mix, and primers, and detection and analysis of the reaction products.

Gene Sequence Analysis Modified from MIT Hypertextbook.

The sequence of DNA base pairs can be analyzed by

- Restriction mapping. Construct a "road map" of restriction sites. A program to do this is WebCutter.
- DNA Sequence Analysis. Cut and clone various restriction fragments, and determine the exact sequence of base pairs. All sequence information is deposited in GenBank. If you just want the sequence of the peptide translated from the RNA, you have to look for insulin mRNA or cDNA.

Once we have a piece of DNA cloned, it is amplified (available in many copies) and we now have a living clone which provides, in theory, an indefinite source of the DNA sequence. How do we analyze the sequence?

DNA Sequence Determination

Once you have identified a particular region of DNA of interest, you need to find out the precise sequence of DNA nucleotides. This is done by di-deoxy sequencing, in which a DNA polymerase is put together with dNTPs in four different reactions, each containing a small amount of one di-deoxy NTP (ATP, TTP, CTP, or GTP). The di-deoxy nucleotide lacks a 3'OH to continue chain extension, so the chain terminates. Each reaction produces a population of fragments terminated at A, T, C, or G.

- The fragments are either radiolabeled or enzymatically labeled. They can be separated on a gel, or on a fluorescence analyzer. All published DNA sequences in the world are deposited in GenBank.

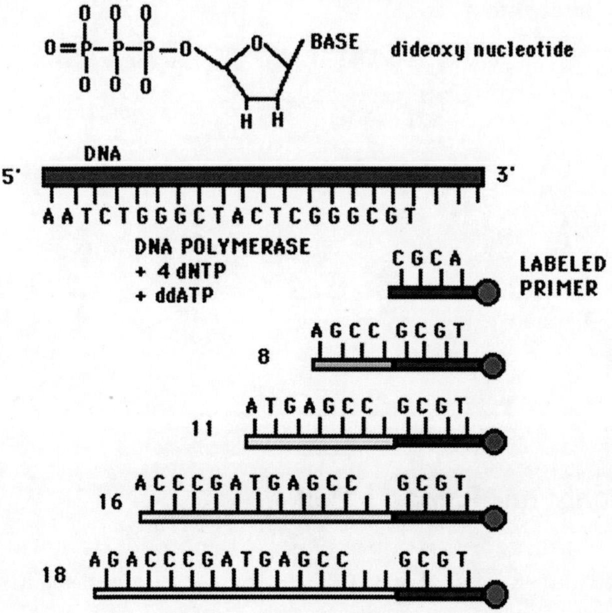

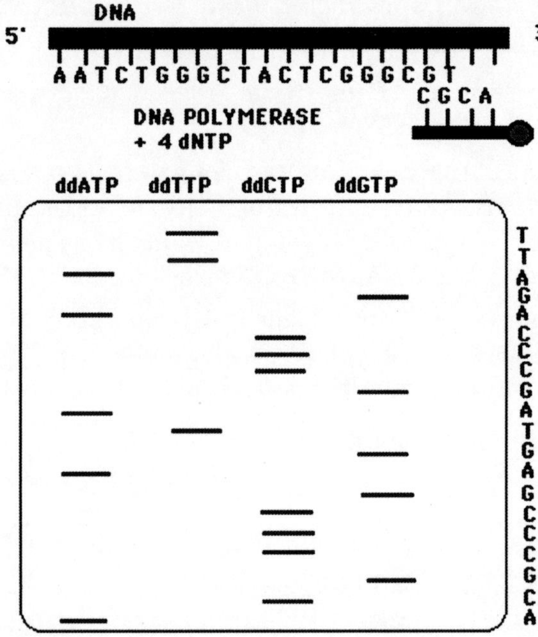

Historically, the products of dideoxy sequencing reactions were subjected to electrophoresis in four different lanes of a gel. Each lane contained a reaction using a different dideoxy terminator nucleotide. The data looked like the image at right.

(c) Different-length strands can be lined up by size to determine DNA sequence.

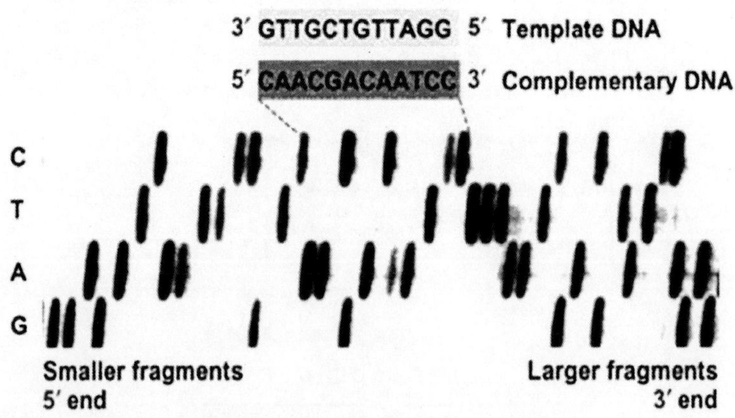

Complementarity and Hybridization

How do we find genes of interest in a genome? Or a particular mRNA in the total cellular RNA? Or a particular protein out of all cell protein?

In solution, hybrid molecular complexes (usually called hybrids) of the following types can exist (other combinations are possible):

- DNA-DNA. A single-stranded DNA molecule (ssDNA probe) can form a double-stranded, base-paired hybrid with a ssDNA target if the probe sequence is the reverse complement of the target sequence. A radiolabeled DNA probe can be applied to DNA from a gel transferred to a membrane, called a Southern Blot (named for its inventor).
- DNA-RNA. A single-stranded DNA (ssDNA) probe molecule can form a double-stranded, base-paired hybrid with an RNA (RNA is usually a single-strand) target if the probe sequence is the reverse complement of the target sequence. An RNA can be radiolabeled to probe a Southern Blot; or, a ssDNA probe can be applied to membrane-bound RNA, called a Northern Blot (name is a pun on Southern.)
- Protein-Protein. An antibody probe molecule (antibodies are proteins) can form a complex with a target protein molecule if the antibody's antigen-binding site can bind to an epitope (small antigenic region) on the target protein. In this case, the hybrid is called an 'antigen-antibody complex' or 'complex' for short. A radiolabeled antibody can probe membrane-bound proteins, called a Western Blot (an even worse pun.)

There are two important features of hybridization:

- Hybridization reactions are specific - the probes will only bind to targets with complimentary sequence (or, in the case of antibodies, sites with the correct 3-d shape).
- Hybridization reactions will occur in the presence of large quantities of molecules similar but not identical to the target. That is, a probe can find one molecule of target in a mixture of zillions of related but non-complementary molecules.

These properties allow you to use hybridization to perform a molecular search for one DNA molecule, or one RNA molecule, or one protein molecule in a complex mixture containing many similar molecules.

Southern, Northern, and Western Blots. Link to summary.

Blots are named for the target molecule.

Southern Blot—DNA cut with restriction enzymes - probed with radioactive DNA.

Northern Blot—RNA - probed with radioactive DNA or RNA.

Example—used to measure angiopoietin angiopoietin expression from cDNA in transgenic mouse.

Western Blot—Protein - probed with radioactive or enzymatically-tagged antibodies. These molecules must then be immobilized on a solid support, so that they will remain in position during probing and washing. The probe is then added, the non-specifically bound probe is removed, and the probe is detected. The place where the probe is detected corresponds to the location of the immobilized target molecule. This process is diagrammed below:

Preparing for Blots

- Southern Blots. DNA is first cut with restriction enzymes and the resulting double-stranded DNA fragments have an extended rod conformation without pre-treatment.
- Northern Blots. Although RNA is single-stranded, RNA molecules often have small regions that can form base-paired secondary structures. To prevent this, the RNA is pre-treated with formaldehyde.
- Western Blots. Proteins have extensive 2' and 3' structures and are not always negatively charged. Proteins are treated with the detergent SDS (sodium dodecyl sulfate) which removes 2' and 3' structure and coats the protein with negative charges.

Transfer to Solid Support. After the DNA, RNA, or protein has been separated by molecular weight, it must be transferred to a solid support before hybridization. (Hybridization does not work well in a gel.) This transfer process is called blotting and is why these hybridization techniques are called blots.

Usually, the solid support is a sheet of nitrocellulose paper (sometimes called a filter because the sheets of nitrocellulose were originally used as filter paper), although other materials are sometimes used. DNA, RNA, and protein stick well to nitrocellulose in a sequence-independent manner.After a series of treatment steps, the probe is added. The probe hybridized to the target molecules is visualized either by autoradiography or by enzyme reaction.

Summary. The important properties of the three blotting procedures of DNA analysis:

Table:

	Southern	Northern	Western
What is separated by molecular weight? (target)	DNA cut with restriction enzymes	RNA denatured with formaldehyde	Protein denatured with SDS
Probe	radioactive gene X DNA	radioactive gene X DNA	Antibody against protein X, labeled with radioactivity or enzyme
What do you learn from it?	Restriction map of gene X in chromosome	-how much gene X mRNA is present? -how long is gene X mRNA?	-how much protein X is present? -how big is protein X?

DNA Microarrays

We can now put most of the protein-encoding genes onto a microarray chip, using technology based on the DNA silicon chip industry. The chip can be used to hybridize to cellular RNA, and measure the expression rates of a large number of genes in a cell.

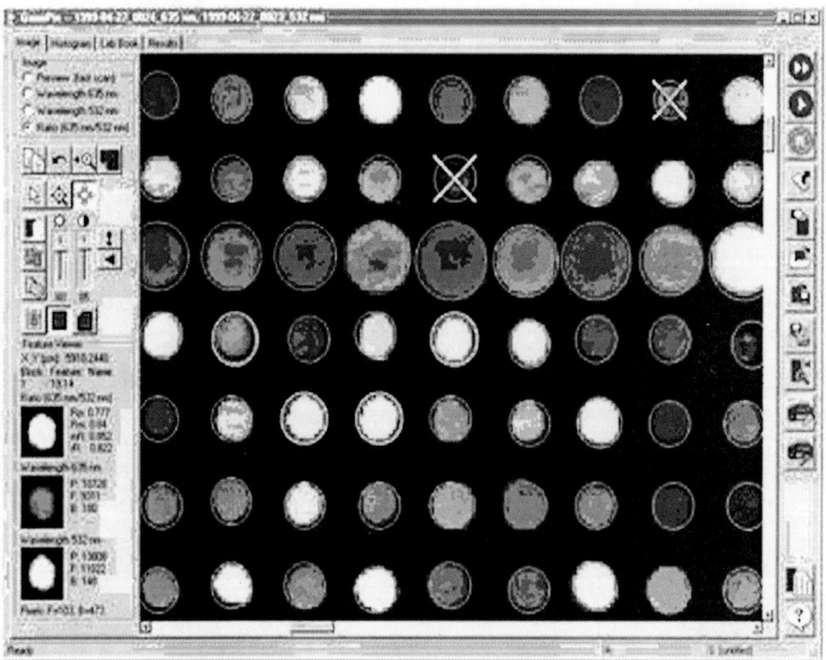

Figure: *Axon Industries.*

Transgenics

Cloning in Animals.

Animals have two different classes of cells: germ line and somatic.

Only alterations in the germ cells can be transmitted to future generations. However, some forms of somatic cell gene therapy can be useful in treating patients.

For example, people with cystic fibrosis can receive the cloned CFTR gene in a nasal spray, which then infects the lining of their lungs and improves lung function. The infected (transformed) epithelial cells eventually are lost, however, during normal processes of tissue growth, so the treatment needs to be repeated. Germ-line cloning. There are two basic ways to clone a gene in the mammalian germ line.

- Inject a DNA fragment containing your gene into the nucleus of a fertilized egg (germ-line gene cloning) or of body tissues in a mature host (somatic gene cloning). The DNA gets taken up at random somewhere in one of the chromosomes.

An example of germ-line cloning is the injection of angiopoietin cDNA into the fertilized mouse egg.

(An example of somatic cloning is gene therapy for cystic fibrosis: inhale a vector containing the CF gene, which gets incorporated into the DNA of cells lining the lungs. Somatic cloned genes are not inherited by offspring.)

- Put a transgene containing positive and negative selective markers into an embryonic stem cell tissue culture. The ES cells take up the gene by homologous recombination, replacing the host allele. The ES cells now are injected into a blastula of an unrelated host, and some of the next generation progeny arise from the ES cells. Once you have cloned a valuable gene into an animal, say a sheep that makes insulin in its milk, how can we produce lots of identical progeny quickly without sexual reproduction? This is what the popular press calls "cloning

But when you create a clone, how do you know it worked? You need to know:

- DNA: Did the transgene really incorporate into the genome? Where?
- RNA: Is the RNA expressed? (Or prevented from expression, by a null allele?)
- Protein: Is a desired protein expressed?

Cloning in Plants

Plants don't have separate somatic and germ cells, so creating transgenic plants is easier than creating transgenic animals. Techniques include the use of microprojectile bombardment (the "gene gun") and the use of Agrobacterium tumefaciens.

3

Molecular Biology Applications in Clonning

Molecular cloning refers to a set of experimental methods in molecular biology that are used to assemble recombinant DNA molecules and to direct their replication within host organisms. The use of the word *cloning* refers to the fact that the method involves the replication of a single DNA molecule starting from a single living cell to generate a large population of cells containing identical DNA molecules. Molecular cloning generally uses DNA sequences from two different organisms: the species that is the source of the DNA to be cloned, and the species that will serve as the living host for replication of the recombinant DNA. Molecular cloning methods are central to many contemporary areas of modern biology and medicine.

In a conventional molecular cloning experiment, the DNA to be cloned is obtained from an organism of interest, then treated with enzymes in the test tube to generate smaller DNA fragments. Subsequently, these fragments are then combined with vector DNA to generate recombinant DNA molecules.

The recombinant DNA is then introduced into a host organism (typically an easy-to-grow, benign, laboratory strain of *E. coli* bacteria). This will generate a population of organisms in which recombinant DNA molecules are replicated along with the host DNA. Because they contain foreign DNA fragments, these are transgenic or genetically-modified microorganisms (GMO). This process takes advantage of the fact that a single bacterial cell can be induced to take up and replicate a single recombinant DNA molecule. This single cell can then be expanded exponentially to generate a large amount of bacteria, each of which contain copies of the original recombinant molecule. Thus, both the resulting bacterial population, and the recombinant DNA molecule, are commonly referred to as "clones". Strictly speaking, *recombinant DNA* refers to DNA molecules, while *molecular cloning* refers to the experimental methods used to assemble them.

History of Molecular Cloning

Prior to the 1970s, our understanding of genetics and molecular biology was severely hampered by an inability to isolate and study individual genes from complex organisms. This changed dramatically with the advent of molecular cloning methods. Microbiologists, seeking to understand the molecular mechanisms through which bacteria restricted the growth of bacteriophage, isolated restriction endonucleases, enzymes that could cleave DNA molecules only when specific DNA sequences were encountered. They showed that restriction enzymes cleaved chromosome-length DNA molecules at specific locations, and that specific sections of the larger molecule could be purified by size fractionation.

Using a second enzyme, DNA ligase, fragments generated by restriction enzymes could be joined in new combinations, termed recombinant DNA. By recombining DNA segments of interest with vector DNA, such as bacteriophage or plasmids, which naturally replicate inside bacteria, large quantities of purified recombinant DNA molecules could be produced in bacterial cultures. The first recombinant DNA molecules were generated and studied in 1972.

Molecular cloning takes advantage of the fact that the chemical structure of DNA is fundamentally the same in all living organisms. Therefore, if any segment of DNA from any organism is inserted into a DNA segment containing the molecular sequences required for DNA replication, and the resulting recombinant DNA is introduced into the organism from which the replication sequences were obtained, then the foreign DNA will be replicated along with the host cell's DNA in the transgenic organism.

Molecular cloning is similar to polymerase chain reaction (PCR) in that it permits the replication of a specific DNA sequence. The fundamental difference between the two methods is that molecular cloning involves replication of the DNA in a living microorganism, while PCR replicates DNA in an *in vitro* solution, free of living cells.

Steps in Molecular Cloning

In standard molecular cloning experiments, the cloning of any DNA fragment essentially involves seven steps: (1) Choice of host organism and cloning vector, (2) Preparation of vector DNA, (3) Preparation of DNA to be cloned, (4) Creation of recombinant DNA, (5) Introduction of recombinant DNA into host organism, (6) Selection of organisms containing recombinant DNA, (7) Screening for clones with desired DNA inserts and biological properties.

Choice of Host Organism and Cloning Vector

Although a very large number of host organisms and molecular cloning vectors are in use, the great majority of molecular cloning experiments begin with a laboratory strain of the bacterium *E. coli* (Escherichia coli) and a plasmid cloning vector.

E. coli and plasmid vectors are in common use because they are technically sophisticated, versatile, widely available, and offer rapid growth of recombinant organisms with minimal equipment. If the DNA to be cloned is exceptionally large (hundreds of thousands to millions of base pairs), then a bacterial artificial chromosome or yeast artificial chromosome vector is often chosen.

Specialised applications may call for specialised host-vector systems. For example, if the experimentalists wish to harvest a particular protein from the recombinant organism, then an expression vector is chosen that contains appropriate signals for transcription and translation in the desired host organism.

Alternatively, if replication of the DNA in different species is desired (for example transfer of DNA from bacteria to plants), then a multiple host range vector (also termed shuttle vector) may be selected.

In practice, however, specialised molecular cloning experiments usually begin with cloning into a bacterial plasmid, followed by subcloning into a specialised vector.

Whatever combination of host and vector are used, the vector almost always contains four DNA segments that are critically important to its function and experimental utility—(1) an origin of DNA replication is necessary for the vector (and recombinant sequences linked to it) to replicate inside the host organism, (2) one or more unique restriction endonuclease recognition sites that serves as sites where foreign DNA may be introduced, (3) a selectable genetic marker gene that can be used to enable the survival of cells that have taken up vector sequences, and (4) an additional gene that can be used for screening which cells contain foreign DNA.

Preparation of Vector DNA

The cloning vector is treated with a restriction endonuclease to cleave the DNA at the site where foreign DNA will be inserted. The restriction enzyme is chosen to generate a configuration at the cleavage site that is compatible with that at the ends of the foreign DNA. Typically, this is done by cleaving the vector DNA and foreign DNA with the same restriction enzyme, for example EcoRI.

Most modern vectors contain a variety of convenient cleavage sites that are unique within the vector molecule (so that the vector can only be cleaved at a single site) and is located within a gene (frequently beta-galactosidase) whose inactivation can be used to distinguish recombinant from non-recombinant organisms at a later step in the process. To improve the ratio of recombinant to non-recombinant organisms, the cleaved vector may be treated with an enzyme (alkaline phosphatase) that modifies the vector ends in such a way that it cannot replicate within cells unless it contains foreign DNA.

Preparation of DNA to be Cloned

For cloning of genomic DNA, the DNA to be cloned is extracted from the organism of interest. Virtually any tissue source can be used (even tissues from extinct animals), as long as the DNA is not extensively degraded. The DNA is then purified using simple methods to remove contaminating proteins (extraction with phenol), RNA (ribonuclease) and smaller molecules (precipitation and/or chromatography). Polymerase chain reaction (PCR) methods are often used for amplification of specific DNA or RNA (RT-PCR) sequences prior to molecular cloning.

DNA for cloning experiments may also be obtained from RNA using reverse transcriptase (complementary DNA or cDNA cloning), or in the form of synthetic DNA (artificial gene synthesis). cDNA cloning is usually used to obtain clones representative of the mRNA population of the cells of interest, while synthetic DNA is used to obtain any precise sequence defined by the designer.

The purified DNA is then treated with a restriction enzyme to generate fragments with ends capable of being linked to those of the vector. If necessary, short double-stranded segments of DNA containing desired restriction sites may be added to create end structures that are compatible with the vector.

Creation of Recombinant DNA with DNA Ligase

The creation of recombinant DNA is in many ways the simplest step of the molecular cloning process. DNA prepared from the vector and foreign source are simply mixed together at appropriate concentrations and exposed to an enzyme (DNA ligase) that covalently links the ends together.

This joining reaction is often termed ligation. The resulting DNA mixture containing randomly joined ends is then ready for introduction into the host organism. DNA ligase only recognises and acts on the ends of linear DNA molecules, usually resulting a complex mixture of DNA molecules with randomly joined ends. The desired products (vector DNA covalently linked to foreign DNA) will be present, but other sequences (e.g. foreign DNA linked to itself, vector DNA linked to itself and higher-order combinations of vector and foreign DNA) are also usually present. This complex mixture is sorted out in subsequent steps of the cloning process, after the DNA mixture is introduced into cells.

Introduction of Recombinant DNA into Host Organism

The DNA mixture, previously manipulated in vitro, is moved back into a living cell, referred to as the host organism. The methods used to get DNA into cells are varied, and the name applied to this step in the molecular cloning process will often depend upon the experimental method that is chosen (e.g. transformation, transduction, transfection, electroporation).

When microorganisms are able to take up and replicate DNA from their local environment, the process is termed transformation, and cells that are in a physiological state such that they can take up DNA are said to be competent. In mammalian cell culture, the analogous process of introducing DNA into cells is commonly termed transfection. Both transformation and transfection usually require preparation of the cells through a special growth regime and chemical treatment process that will vary with the specific species and cell types that are used.

Electroporation uses high voltage electrical pulses to translocate DNA across the cell membrane (and cell wall, if present). In contrast, transduction involves the packaging of DNA into virus-derived particles, and using these virus-like particles to introduce the encapsulated DNA into the cell through a process resembling viral infection. Although electroporation and transduction are highly specialised methods, they may be the most efficient methods to move DNA into cells.

Selection of Organisms Containing Vector Sequences

Whichever method is used, the introduction of recombinant DNA into the chosen host organism is usually a low efficiency process; that is, only a small fraction of the cells will actually take up DNA. Experimental scientists deal with this issue through a step of artificial genetic selection, in which cells that have not taken up DNA are selectively killed, and only those cells that can actively replicate DNA containing the selectable marker gene encoded by the vector are able to survive.

When bacterial cells are used as host organisms, the selectable marker is usually a gene that confers resistance to an antibiotic that would otherwise kill the cells, typically ampicillin. Cells harbouring the vector will survive when exposed to the antibiotic, while those that have failed to take up vector sequences will die. When mammalian cells (e.g. human or mouse cells) are used, a similar strategy is used, except that the marker gene confers resistance to the antibiotic Geneticin.

Screening for Clones with Desired DNA Inserts and Biological Properties

Modern bacterial cloning vectors (e.g. pUC19 and later derivatives including the pGEM vectors) use the blue-white screening system to distinguish colonies (clones) of transgenic cells from those that contain the parental vector (i.e. vector DNA with no recombinant sequence inserted). In these vectors, foreign DNA is inserted into a sequence that encodes an essential part of beta-galactosidase, an enzyme whose activity results in formation of a blue-coloured colony on the culture medium that is used for this work. Insertion of the foreign DNA into the beta-galactosidase coding sequence disables the function of the enzyme, so that colonies containing recombinant plasmids remain colourless (white). Therefore, experimentalists are easily able to identify and conduct further studies on transgenic bacterial clones, while ignoring those that do not contain recombinant DNA.

The total population of individual clones obtained in a molecular cloning experiment is often termed a DNA library. Libraries may be highly complex (as when cloning complete genomic DNA from an organism) or relatively simple (as when moving a previously-cloned DNA fragment into a different plasmid), but it is almost always necessary to examine a number of different clones to be sure that the desired DNA construct is obtained. This may be accomplished through a very wide range of experimental methods, including the use of nucleic acid hybridisations, antibody probes, polymerase chain reaction, restriction fragment analysis and/or DNA sequencing.

Applications of Molecular Cloning

Molecular cloning provides scientists with an essentially unlimited quantity of any individual DNA segments derived from any genome. This material can be used for a wide range of purposes, including those in both basic and applied biological science. A few of the more important applications are summarised here.

Genome Organisation and Gene Expression

Molecular cloning has led directly to the elucidation of the complete of sexual DNA sequence of the genomes of a very large number of species and to an exploration of genetic diversity within individual species, work that has been done mostly by determining the DNA sequence of large numbers of randomly cloned fragments of the genome, and assembling the overlapping sequences.

At the level of individual genes, molecular clones are used to generate probes that are used for examining how genes are expressed, and how that expression is related to other processes in biology, including the metabolic environment, extracellular signals, development, learning, senescence and cell death. Cloned genes can also provide tools to examine the biological function and importance of individual genes, by allowing investigators to inactivate the genes, or make more subtle mutations using regional mutagenesis or site-directed mutagenesis.

Production of Recombinant Proteins

Obtaining the molecular clone of a gene can lead to the development of organisms that produce the protein product of the cloned genes, termed a recombinant protein. In practice, it is frequently more difficult to develop an organism that produces an active form of the recombinant protein in desirable quantities than it is to clone the gene. This is because the molecular signals for gene expression are complex and variable, and because protein folding, stability and transport can be very challenging.

Many useful proteins are currently available as recombinant products. These include—(1) medically-useful proteins whose administration can correct a defective or poorly-expressed gene (e.g. recombinant factor VIII, a blood-clotting factor deficient in some forms of hemophilia, and recombinant insulin, used to treat some forms of diabetes), (2) proteins that can be administered to assist in a life threatening emergency (e.g. tissue plasminogen activator, used to treat strokes), and (3) recombinant subunit vaccines, in which a purified protein can be used to immunize patients against infectious diseases, without exposing them to the infectious agent itself (e.g. hepatitis B vaccine).

Transgenic Organisms

Once characterised and manipulated to provide signals for appropriate expression, cloned genes may be inserted into organisms, generating transgenic organisms, also termed genetically-modified organisms (GMOs). Although most GMOs are generated for purposes of basic biological research, a number of GMOs have been developed for commercial use, ranging from animals and plants that produce pharmaceuticals or other compounds (pharming), herbicide-resistant crop plants, and fluorescent tropical fish (GloFish) for home entertainment.

Gene Therapy

Gene therapy involves supplying a functional gene to cells lacking that function, with the aim of correcting a genetic disorder or acquired disease. Gene therapy can be broadly divided into two categories. The first is alteration of germ cells, that is, sperm or eggs, which results in a permanent genetic change for the whole organism and subsequent generations. This "germ line gene therapy" is considered by many to be unethical in human beings. The second type of gene therapy, "somatic cell gene therapy", is analogous to an organ transplant. In this case, one or more specific tissues are targeted by direct treatment or by removal of the tissue, addition of the therapeutic gene or genes in the laboratory, and return of the treated cells to the patient. Clinical trials of somatic cell gene therapy began in the late 1990s, mostly for the treatment of cancers and blood, liver, and lung disorders.

Despite a great deal of publicity and promises, the history of human gene therapy has been characterised by relatively limited success. The effect of introducing a gene into cells often promotes only partial and/or transient relief from the symptoms of the disease being treated. Some gene therapy trial patients have suffered adverse consequences of the treatment itself, including deaths. In some cases, the adverse effects result from disruption of essential genes within the patient's genome by insertional inactivation. In others, viral vectors used for gene therapy have been contaminated with infectious virus. Nevertheless, gene therapy is still held to be a promising future area of medicine, and is an area where there is a significant level of research and development activity.

Cloning of Viral Antigens

A virus is a small infectious agent that can replicate only inside the living cells of organisms. Viruses infect all types of organisms, from animals and plants to bacteria and archaea. Since Dmitri Ivanovsky's 1892 article describing a non-bacterial pathogen infecting tobacco plants, and the discovery of the tobacco mosaic virus by Martinus Beijerinck in 1898, about 5,000 viruses have been described in detail, although there are millions of different types. Viruses are found in almost every ecosystem on Earth and are the most abundant type of biological entity. The study of viruses is known as virology, a sub-speciality of microbiology.

Virus particles (known as *virions*) consist of two or three parts: the genetic material made from either DNA or RNA, long molecules that carry genetic information; a protein coat that protects these genes; and in some cases an envelope of lipids that surrounds the protein coat when they are outside a cell. The shapes of viruses range from simple helical and icosahedral forms to more complex structures. The average virus is about one one-hundredth the size of the average bacterium. Most viruses are too small to be seen directly with a light microscope.

The origins of viruses in the evolutionary history of life are unclear: some may have evolved from plasmids – pieces of DNA that can move between cells – while others may have evolved from bacteria. In evolution, viruses are an important means of horizontal gene transfer, which increases genetic diversity.

Viruses spread in many ways; viruses in plants are often transmitted from plant to plant by insects that feed on the sap of plants, such as aphids; viruses in animals can be carried by blood-sucking insects. These disease-bearing organisms are known as vectors. Influenza viruses are spread by coughing and sneezing. Norovirus and rotavirus, common causes of viral gastroenteritis, are transmitted by the faecal-oral route and are passed from person to person by contact, entering the body in food or water. HIV is one of several viruses transmitted through sexual contact and by exposure to infected blood. The range of host cells that a virus can infect is called its "host range". This can be narrow or, as when a virus is capable of infecting many species, broad.

Viral infections in animals provoke an immune response that usually eliminates the infecting virus. Immune responses can also be produced by vaccines, which confer an artificially acquired immunity to the specific viral infection. However, some viruses including those causing AIDS and viral hepatitis evade these immune responses and result in chronic infections. Antibiotics have no effect on viruses, but several antiviral drugs have been developed.

Etymology

The word is from the Latin *virus* referring to poison and other noxious substances, first used in English in 1392. *Virulent*, from Latin *virulentus* (poisonous), dates to 1400. A meaning of "agent that causes infectious disease" is first recorded in 1728, before the discovery of viruses by Dmitry Ivanovsky in 1892. The plural is *viruses*. The adjective *viral* dates to 1948. The term *virion* is also used to refer to a single infective viral particle.

History

Louis Pasteur was unable to find a causative agent for rabies and speculated about a pathogen too small to be detected using a microscope. In 1884, the French microbiologist Charles Chamberland invented a filter (known today as the Chamberland filter or Chamberland-Pasteur filter) with pores smaller than bacteria. Thus, he could pass a solution containing bacteria through the filter and completely remove them from the solution.

In 1892, the Russian biologist Dmitry Ivanovsky used this filter to study what is now known as the tobacco mosaic virus. His experiments showed that crushed leaf extracts from infected tobacco plants remain infectious after filtration. Ivanovsky suggested the infection might be caused by a toxin produced by bacteria, but did not pursue the idea. At the time it was thought that all infectious agents could be retained by filters and grown on a nutrient medium – this was part of the germ theory of disease. In 1898, the Dutch microbiologist Martinus Beijerinck repeated the experiments and became convinced that the filtered solution contained a new form of infectious agent. He observed that the agent multiplied only in cells that were dividing, but as his experiments did not show that it was made of particles, he called it a *contagium vivum fluidum* (soluble living germ) and reintroduced the word *virus*.

Beijerinck maintained that viruses were liquid in nature, a theory later discredited by Wendell Stanley, who proved they were particulate. In the same year Friedrich Loeffler and Frosch passed the first animal virus – agent of foot-and-mouth disease (aphthovirus) – through a similar filter.

In the early 20th century, the English bacteriologist Frederick Twort discovered a group of viruses that infect bacteria, now called bacteriophages (or commonly *phages*), and the French-Canadian microbiologist Félix d'Herelle described viruses that, when added to bacteria on agar, would produce areas of dead bacteria. He accurately diluted a suspension of these viruses and discovered that the highest dilutions (lowest virus concentrations), rather than killing all the bacteria, formed discrete areas of dead organisms.

Counting these areas and multiplying by the dilution factor allowed him to calculate the number of viruses in the original suspension. Phages were heralded as a potential treatment for diseases such as typhoid and cholera, but their promise was forgotten with the development of penicillin. The study of phages provided insights into the switching on and off of genes, and a useful mechanism for introducing foreign genes into bacteria.

By the end of the 19th century, viruses were defined in terms of their infectivity, their ability to be filtered, and their requirement for living hosts. Viruses had been grown only in plants and animals. In 1906, Ross Granville Harrison invented a method for growing tissue in lymph, and, in 1913, E. Steinhardt, C. Israeli, and R. A. Lambert used this method to grow vaccinia virus in fragments of guinea pig corneal tissue. In 1928, H. B. Maitland and M. C. Maitland grew vaccinia virus in suspensions of minced hens' kidneys. Their method was not widely adopted until the 1950s, when poliovirus was grown on a large scale for vaccine production. Another breakthrough came in 1931, when the American pathologist Ernest William Goodpasture grew influenza and several other viruses in fertilized chickens' eggs. In 1949, John F. Enders, Thomas Weller, and Frederick Robbins grew polio virus in cultured human embryo cells, the first virus to be grown without using solid animal tissue or eggs. This work enabled Jonas Salk to make an effective polio vaccine.

The first images of viruses were obtained upon the invention of electron microscopy in 1931 by the German engineers Ernst Ruska and Max Knoll. In 1935, American biochemist and virologist Wendell Meredith Stanley examined the tobacco mosaic virus and found it was mostly made of protein. A short time later, this virus was separated into protein and RNA parts. The tobacco mosaic virus was the first to be crystallised and its structure could therefore be elucidated in detail. The first X-ray diffraction pictures of the crystallised virus were obtained by Bernal and Fankuchen in 1941. On the basis of her pictures, Rosalind Franklin discovered the full DNA structure of the virus in 1955. In the same year, Heinz Fraenkel-Conrat and Robley Williams showed that purified tobacco mosaic virus RNA and its coat protein can assemble by themselves to form functional viruses, suggesting that this simple mechanism was probably the means through which viruses were created within their host cells.

The second half of the 20th century was the golden age of virus discovery and most of the 2,000 recognised species of animal, plant, and bacterial viruses were discovered during these years. In 1957, equine arterivirus and the cause of Bovine virus diarrhea (a pestivirus) were discovered. In 1963, the hepatitis B virus was discovered by Baruch Blumberg, and in 1965, Howard Temin described the first retrovirus. Reverse transcriptase, the key enzyme that retroviruses use to translate their RNA into DNA, was first described in 1970, independently by Howard Martin Temin and David Baltimore. In 1983 Luc Montagnier's team at the Pasteur Institute in France, first isolated the retrovirus now called HIV.

Origins

Viruses are found wherever there is life and have probably existed since living cells first evolved. The origin of viruses is unclear because they do not form fossils, so molecular techniques have been used to compare the DNA or RNA of viruses and are a useful means of investigating how they arose. There are three main hypotheses that try to explain the origins of viruses:

Regressive hypothesis : Viruses may have once been small cells that parasitised larger cells. Over time, genes not required by their parasitism were lost. The bacteria rickettsia and chlamydia are living cells that, like viruses, can reproduce only inside host cells. They lend support to this hypothesis, as their dependence on parasitism is likely to have caused the loss of genes that enabled them to survive outside a cell. This is also called the *degeneracy hypothesis*, or *reduction hypothesis*.

Cellular origin hypothesis: Some viruses may have evolved from bits of DNA or RNA that "escaped" from the genes of a larger organism. The escaped DNA could have come from plasmids (pieces of naked DNA that can move *between* cells) or transposons (molecules of DNA that replicate and move around to different positions *within* the genes of the cell). Once called "jumping genes", transposons are examples of mobile genetic elements and could be the origin of some viruses. They were discovered in maize by Barbara McClintock in 1950. This is sometimes called the *vagrancy hypothesis*, or the *escape hypothesis*.

Coevolution hypothesis: This is also called the *virus-first hypothesis* and proposes that viruses may have evolved from complex molecules of protein and nucleic acid at the same time as cells first appeared on earth and would have been dependent on cellular life for billions of years. Viroids are molecules of RNA that are not classified as viruses because they lack a protein coat.

However, they have characteristics that are common to several viruses and are often called subviral agents. Viroids are important pathogens of plants. They do not code for proteins but interact with the host cell and use the host machinery for their replication. The hepatitis delta virus of humans has an RNA genome similar to viroids but has a protein coat derived from hepatitis B virus and cannot produce one of its own. It is, therefore, a defective virus and cannot replicate without the help of hepatitis B virus. In similar manner, the virophage 'sputnik' is dependent on mimivirus, which infects the protozoan *Acanthamoeba castellanii*. These viruses that are dependent on the presence of other virus species in the host cell are called *satellites* and may represent evolutionary intermediates of viroids and viruses.

In the past, there were problems with all of these hypotheses: the regressive hypothesis did not explain why even the smallest of cellular parasites do not resemble viruses in any way. The escape hypothesis did not explain the complex capsids and other structures on virus particles. The virus-first hypothesis contravened the definition of viruses in that they require host cells. Viruses are now recognised as ancient and to have origins that predate the divergence of life into the three domains. This discovery has led modern virologists to reconsider and re-evaluate these three classical hypotheses.

The evidence for an ancestral world of RNA cells and computer analysis of viral and host DNA sequences are giving a better understanding of the evolutionary relationships between different viruses and may help identify the ancestors of modern viruses. To date, such analyses have not proved which of these hypotheses is correct. However, it seems unlikely that all currently known viruses have a common ancestor, and viruses have probably arisen numerous times in the past by one or more mechanisms.

Prions are infectious protein molecules that do not contain DNA or RNA. They cause an infection in sheep called scrapie and cattle bovine spongiform encephalopathy ("mad cow" disease). In humans they cause kuru and Creutzfeldt-Jakob disease. They are able to replicate because some proteins can exist in two different shapes and the prion changes the normal shape of a host protein into the prion shape. This starts a chain reaction where each prion protein converts many host proteins into more prions, and these new prions then go on to convert even more protein into prions. Although they are fundamentally different from viruses and viroids, their discovery gives credence to the idea that viruses could have evolved from self-replicating molecules.

Microbiology

Life Properties

Opinions differ on whether viruses are a form of life, or organic structures that interact with living organisms. They have been described as "organisms at the edge of life", since they resemble organisms in that they possess genes and evolve by natural selection, and reproduce by creating multiple copies of themselves through self-assembly.

Although they have genes, they do not have a cellular structure, which is often seen as the basic unit of life. Viruses do not have their own metabolism, and require a host cell to make new products. They therefore cannot naturally reproduce outside a host cell – although bacterial species such as rickettsia and chlamydia are considered living organisms despite the same limitation.

Accepted forms of life use cell division to reproduce, whereas viruses spontaneously assemble within cells. They differ from autonomous growth of crystals as they inherit genetic mutations while being subject to natural selection. Virus self-assembly within host cells has implications for the study of the origin of life, as it lends further credence to the hypothesis that life could have started as self-assembling organic molecules.

Structure

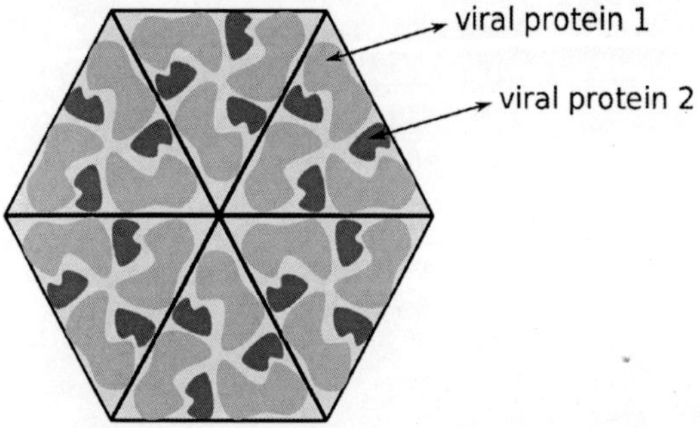

Hexon of a virus capsid made from two protein molecules

Figure : *Diagram of how a virus capsid can be constructed using multiple copies of just two protein molecules*

Viruses display a wide diversity of shapes and sizes, called *morphologies*. Generally viruses are much smaller than bacteria. Most viruses that have been studied have a diameter between 20 and 300 nanometres. Some filoviruses have a total length of up to 1400 nm; their diameters are only about 80 nm. Most viruses cannot be seen with a light microscope so scanning and transmission electron microscopes are used to visualise virions. To increase the contrast between viruses and the background, electron-dense "stains" are used. These are solutions of salts of heavy metals, such as tungsten, that scatter the electrons from regions covered with the stain. When virions are coated with stain (positive staining), fine detail is obscured. Negative staining overcomes this problem by staining the background only.

A complete virus particle, known as a virion, consists of nucleic acid surrounded by a protective coat of protein called a capsid. These are formed from identical protein subunits called capsomeres. Viruses can have a lipid "envelope" derived from the host cell membrane. The capsid is made from proteins encoded by the viral genome and its shape serves as the basis for morphological distinction. Virally coded protein subunits will self-assemble to form a capsid, generally requiring the presence of the virus genome. Complex viruses code for proteins that assist in the construction of their capsid. Proteins associated with nucleic acid are known as nucleoproteins, and the association of viral capsid proteins with viral nucleic acid is called a nucleocapsid. The capsid and entire virus structure can be mechanically (physically) probed through atomic force microscopy. In general, there are four main morphological virus types:

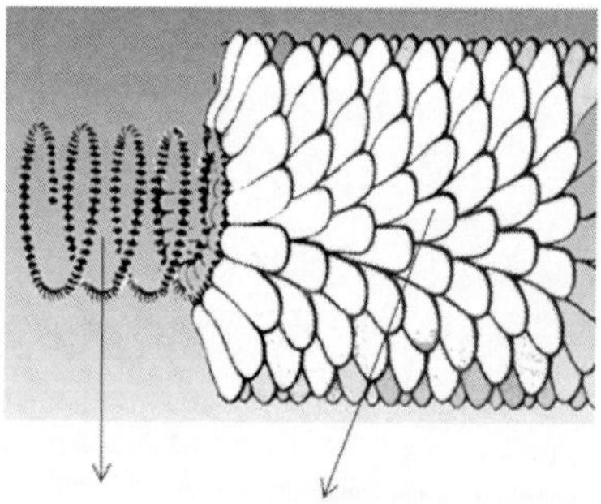

Figure : *RNA coiled in a helix of repeating protein subunits*

Helical: These viruses are composed of a single type of capsomer stacked around a central axis to form a helical structure, which may have a central cavity, or hollow tube. This arrangement results in rod-shaped or filamentous virions: These can be short and highly rigid, or long and very flexible.

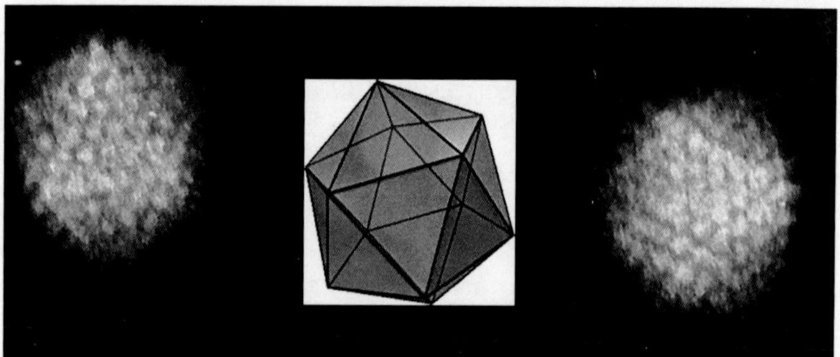

Figure : *Electron micrograph of icosahedral adenovirus*

The genetic material, in general, single-stranded RNA, but ssDNA in some cases, is bound into the protein helix by interactions between the negatively charged nucleic acid and positive charges on the protein. Overall, the length of a helical capsid is related to the length of the nucleic acid contained within it and the diameter is dependent on the size and arrangement of capsomers. The well-studied tobacco mosaic virus is an example of a helical virus.

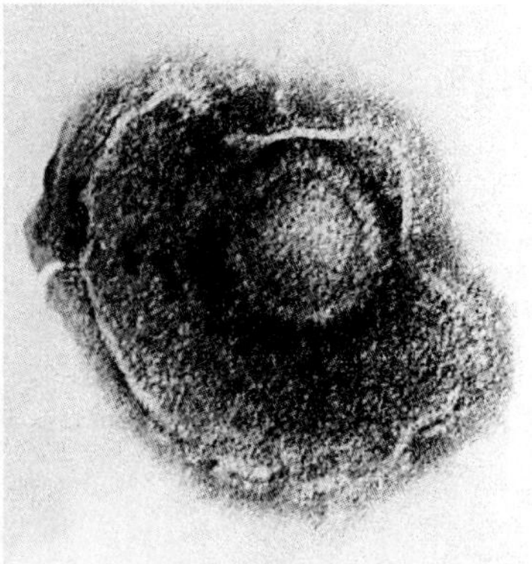

Figure : *Herpes viruses have a lipid envelope*

Icosahedral: Most animal viruses are icosahedral or near-spherical with icosahedral symmetry. A regular icosahedron is the optimum way of forming a closed shell from identical subunits. The minimum number of identical capsomers required is twelve, each composed of five identical subunits.

Many viruses, such as rotavirus, have more than twelve capsomers and appear spherical but they retain this symmetry. Capsomers at the apices are surrounded by five other capsomers and are called pentons. Capsomers on the triangular faces are surrounded by six others and are called hexons. Hexons are essentially flat and pentons, which form the 12 vertices, are curved. The same protein may act as the subunit of both the pentamers and hexamers or they may be composed of different proteins.

Although the isosahedral structure is extremely common among viruses because of size differences slight variations exist between virions. Given an asymmetric subunit on a triangular face of a regular isosahedron, with three subunits per face 60 such subunits can be placed in an equivalent manner. Most virions, because of their size, have more than 60 subunits. These variations have been classified on the basis of the quasi-equivalence principle proposed by Caspar and Klug.

An isosahedral structure can be regarded as being constructed from 12 pentamers. The number of pentamers is fixed but the number of hexamers can vary. These shells can be constructed from pentamers and hexamers by minimising the number T (triangulation number) of nonequivalent locations that subunits occupy, with the T-number adopting the particular integer values 1, 3, 4, 7, 12, 13,...($T = h^2 + k^2 + hk$, with h, k equal to nonnegative integers). These shells always contain 12 pentamers plus 10 (T-1) hexamers. Although this classification can be applied to the majority of known viruses exceptions are known including the retroviruses where point mutations disrupt the symmetry.

Prolate: This is an isosahedron elongated along the fivefold axis and is a common arrangement of the heads of bacteriophages. Such a structure is composed of a cylinder with a cap at either end. The cylinder is composed of 10 triangles. The Q number, which can be can be any positive integer, specifies the number of triangles, composed of asymmetric subunits, that make up the 10 triangles of the cylinder. The caps are classified by the T number.

Envelope: Some species of virus envelop themselves in a modified form of one of the cell membranes, either the outer membrane surrounding an infected host cell or internal membranes such as nuclear membrane or endoplasmic reticulum, thus gaining an outer lipid bilayer known as a viral envelope. This membrane is studded with proteins coded for by the viral genome and host genome; the lipid membrane itself and any carbohydrates present originate entirely from the host. The influenza virus and HIV use this strategy. Most enveloped viruses are dependent on the envelope for their infectivity.

Complex: These viruses possess a capsid that is neither purely helical nor purely icosahedral, and that may possess extra structures such as protein tails or a complex outer wall. Some bacteriophages, such as Enterobacteria phage T4, have a complex structure consisting of an icosahedral head bound to a helical tail, which may have a hexagonal base plate with protruding protein tail fibres. This tail structure acts like a molecular syringe, attaching to the bacterial host and then injecting the viral genome into the cell.

The poxviruses are large, complex viruses that have an unusual morphology. The viral genome is associated with proteins within a central disk structure known as a nucleoid. The nucleoid is surrounded by a membrane and two lateral bodies of unknown function. The virus has an outer envelope with a thick layer of protein studded over its surface. The whole virion is slightly pleiomorphic, ranging from ovoid to brick shape. Mimivirus is the largest characterised virus, with a capsid diameter of 400 nm. Protein filaments measuring 100 nm project from the surface. The capsid appears hexagonal under an electron microscope, therefore the capsid is probably icosahedral. In 2011, researchers discovered a larger virus on ocean floor of the coast of Las Cruces, Chile. Provisionally named *Megavirus chilensis*, it can be seen with a basic light microscope.

Some viruses that infect Archaea have complex structures that are unrelated to any other form of virus, with a wide variety of unusual shapes, ranging from spindle-shaped structures, to viruses that resemble hooked rods, teardrops or even bottles. Other archaeal viruses resemble the tailed bacteriophages, and can have multiple tail structures.

Genome

An enormous variety of genomic structures can be seen among viral species; as a group they contain more structural genomic diversity than plants, animals, archaea, or bacteria. There are millions of different types of viruses, although only about 5,000 of them have been described in detail. A virus has either DNA or RNA genes and is called a DNA virus or a RNA virus respectively. The vast majority of viruses have RNA genomes. Plant viruses tend to have single-stranded RNA genomes and bacteriophages tend to have double-stranded DNA genomes.

Viral genomes are circular, as in the polyomaviruses, or linear, as in the adenoviruses. The type of nucleic acid is irrelevant to the shape of the genome. Among RNA viruses and certain DNA viruses, the genome is often divided up into separate parts, in which case it is called *segmented*. For RNA viruses, each segment often codes for only one protein and they are usually found together in one capsid. However, all segments are not required to be in the same virion for the virus to be infectious, as demonstrated by brome mosaic virus and several other plant viruses.

A viral genome, irrespective of nucleic acid type, is almost always either single-stranded or double-stranded. Single-stranded genomes consist of an unpaired nucleic acid, analogous to one-half of a ladder split down the middle. Double-stranded genomes consist of two complementary paired nucleic acids, analogous to a ladder. The virus particles of some virus families, such as those belonging to the *Hepadnaviridae*, contain a genome that is partially double-stranded and partially single-stranded. For most viruses with RNA genomes and some with single-stranded DNA genomes, the single strands are said to be either positive-sense (called the plus-strand) or negative-sense (called the minus-strand), depending on whether or not they are complementary to the viral messenger RNA (mRNA). Positive-sense viral RNA is in the same sense as viral mRNA and thus at least a part of it can be immediately translated by the host cell. Negative-sense viral RNA is complementary to mRNA and thus must be converted to positive-sense RNA by an RNA-dependent RNA polymerase before translation.

DNA nomenclature for viruses with single-sense genomic ssDNA is similar to RNA nomenclature, in that the *coding strand* for the viral mRNA is complementary to it (“), and the *non-coding strand* is a copy of it (+). However, several types of ssDNA and ssRNA viruses have genomes that are ambisense in that transcription can occur off both strands in a double-stranded replicative intermediate. Examples include geminiviruses, which are ssDNA plant viruses and arenaviruses, which are ssRNA viruses of animals.

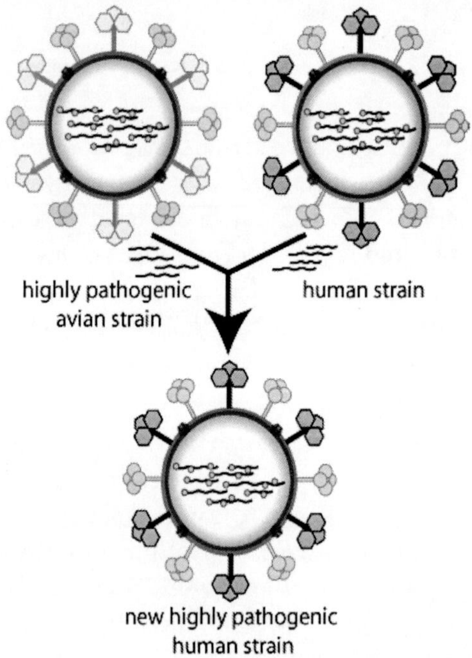

***Figure :** How antigenic shift, or reassortment, can result in novel and highly pathogenic strains of human influenza*

Genome size varies greatly between species. The smallest viral genomes – the ssDNA circoviruses, family *Circoviridae* – code for only two proteins and have a genome size of only 2 kilobases; the largest – mimiviruses – have genome sizes of over 1.2 megabases and code for over one thousand proteins. RNA viruses generally have smaller genome sizes than DNA viruses because of a higher error-rate when replicating, and have a maximum upper size limit. Beyond this limit, errors in the genome when replicating render the virus useless or uncompetitive. To compensate for this, RNA viruses often have segmented genomes – the genome is split into smaller molecules – thus reducing the chance that an error in a single-component genome will incapacitate the entire genome. In contrast, DNA viruses generally have larger genomes because of the high fidelity of their replication enzymes. Single-strand DNA viruses are an exception to this rule, however, as mutation rates for these genomes can approach the extreme of the ssRNA virus case.

Viruses undergo genetic change by several mechanisms. These include a process called genetic drift where individual bases in the DNA or RNA mutate to other bases. Most of these point mutations are "silent" – they do not change the protein that the gene encodes – but others can confer evolutionary advantages such as resistance to antiviral drugs.

Antigenic shift occurs when there is a major change in the genome of the virus. This can be a result of recombination or reassortment. When this happens with influenza viruses, pandemics might result. RNA viruses often exist as quasispecies or swarms of viruses of the same species but with slightly different genome nucleoside sequences. Such quasispecies are a prime target for natural selection.

Segmented genomes confer evolutionary advantages; different strains of a virus with a segmented genome can shuffle and combine genes and produce progeny viruses or (offspring) that have unique characteristics. This is called reassortment or *viral sex*. Genetic recombination is the process by which a strand of DNA is broken and then joined to the end of a different DNA molecule. This can occur when viruses infect cells simultaneously and studies of viral evolution have shown that recombination has been rampant in the species studied. Recombination is common to both RNA and DNA viruses.

Replication Cycle

Viral populations do not grow through cell division, because they are acellular. Instead, they use the machinery and metabolism of a host cell to produce multiple copies of themselves, and they *assemble* in the cell.

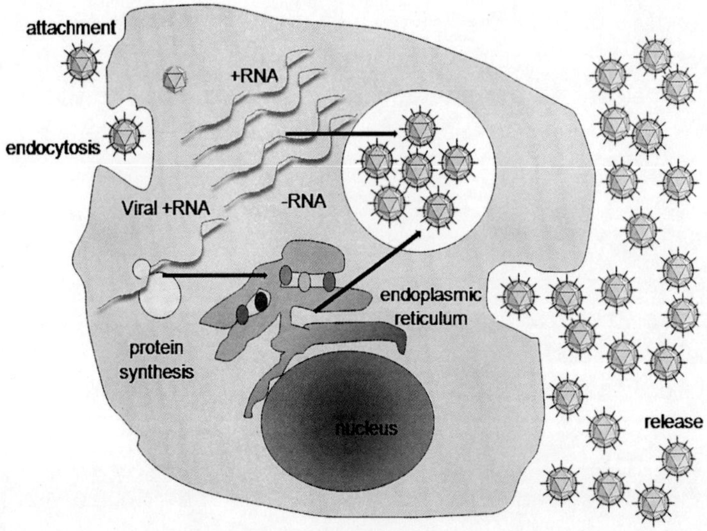

Figure : *A typical virus replication cycle*

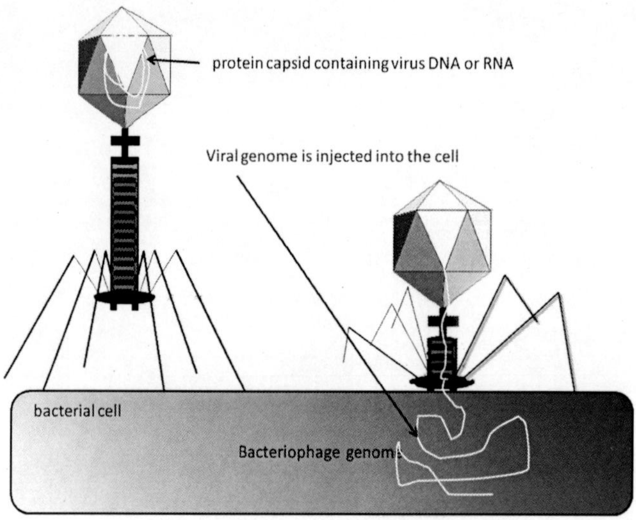

Figure : *Some bacteriophages inject their genomes into bacterial cells*

The life cycle of viruses differs greatly between species but there are six *basic* stages in the life cycle of viruses:

- *Attachment* is a specific binding between viral capsid proteins and specific receptors on the host cellular surface. This specificity determines the host range of a virus. For example, HIV infects a limited range of human leucocytes. This is because its surface protein, gp120, specifically interacts with the CD4 molecule – a chemokine receptor – which is most commonly found on the surface of CD4+ T-Cells. This mechanism has evolved to favour those viruses that infect only cells in which they are capable of replication. Attachment to the receptor can induce the viral envelope protein to undergo changes that results in the fusion of viral and cellular membranes, or changes of non-enveloped virus surface proteins that allow the virus to enter.
- *Penetration* follows attachment: Virions enter the host cell through receptor-mediated endocytosis or membrane fusion. This is often called viral entry. The infection of plant and, it is presumed, fungal cells is different from that of animal cells. Plants have a rigid cell wall made of cellulose, and fungi one of chitin, so most viruses can get inside these cells only after trauma to the cell wall. However, nearly all plant viruses (such as tobacco mosaic virus) can also move directly from cell to cell, in the form of single-stranded nucleoprotein complexes, through pores called plasmodesmata. This process requires movement proteins, which are virus-encoded proteins probably originally derived from plant proteins, which interact with the plasmodesmatal transport machinery Bacteria, like plants, have strong cell walls that a virus must breach

to infect the cell. However, given that bacterial cell walls are much less thick than plant cell walls due to their much smaller size, some viruses have evolved mechanisms that inject their genome into the bacterial cell across the cell wall, while the viral capsid remains outside.

- *Uncoating* is a process in which the viral capsid is removed: This may be by degradation by viral enzymes or host enzymes or by simple dissociation; the end-result is the releasing of the viral genomic nucleic acid.
- *Replication* of viruses primarily involves multiplication of the genome. Replication involves synthesis of viral messenger RNA (mRNA) from "early" genes (with exceptions for positive sense RNA viruses), viral protein synthesis, possible assembly of viral proteins, then viral genome replication mediated by early or regulatory protein expression. This may be followed, for complex viruses with larger genomes, by one or more further rounds of mRNA synthesis: "late" gene expression is, in general, of structural or virion proteins.
- Following the structure-mediated self-*assembly* of the virus particles, some modification of the proteins often occurs. In viruses such as HIV, this modification (sometimes called maturation) occurs *after* the virus has been released from the host cell.
- Viruses can be *released* from the host cell by lysis, a process that kills the cell by bursting its membrane and cell wall if present: This is a feature of many bacterial and some animal viruses. Some viruses undergo a lysogenic cycle where the viral genome is incorporated by genetic recombination into a specific place in the host's chromosome. The viral genome is then known as a "provirus" or, in the case of bacteriophages a "prophage". Whenever the host divides, the viral genome is also replicated. The viral genome is mostly silent within the host; however, at some point, the provirus or prophage may give rise to active virus, which may lyse the host cells. Enveloped viruses (e.g., HIV) typically are released from the host cell by budding. During this process the virus acquires its envelope, which is a modified piece of the host's plasma or other, internal membrane.

The genetic material within virus particles, and the method by which the material is replicated, varies considerably between different types of viruses.

DNA viruses: The genome replication of most DNA viruses takes place in the cell's nucleus. If the cell has the appropriate receptor on its surface, these viruses enter the cell sometimes by direct fusion with the cell membrane (e.g. herpesviruses) or – more usually – by receptor-mediated endocytosis. Most DNA viruses are entirely dependent on the host cell's DNA and RNA synthesising machinery, and RNA processing machinery; however, viruses with larger genomes may encode much of this machinery themselves. In eukaryotes the viral genome must cross the cell's nuclear membrane to access this machinery, while in bacteria it need only enter the cell.

RNA viruses: Replication usually takes place in the cytoplasm. RNA viruses can be placed into four different groups depending on their modes of replication. The polarity (whether or not it can be used directly by ribosomes to make proteins) of single-stranded RNA viruses largely determines the replicative mechanism; the other major criterion is whether the genetic material is single-stranded or double-stranded. All RNA viruses use their own RNA replicase enzymes to create copies of their genomes.

Reverse transcribing viruses: These have ssRNA (*Retroviridae, Metaviridae, Pseudoviridae*) or dsDNA (*Caulimoviridae*, and *Hepadnaviridae*) in their particles. Reverse transcribing viruses with RNA genomes (retroviruses), use a DNA intermediate to replicate, whereas those with DNA genomes (pararetroviruses) use an RNA intermediate during genome replication. Both types use a reverse transcriptase, or RNA-dependent DNA polymerase enzyme, to carry out the nucleic acid conversion. Retroviruses integrate the DNA produced by reverse transcription into the host genome as a provirus as a part of the replication process; pararetroviruses do not, although integrated genome copies of especially plant pararetroviruses can give rise to infectious virus. They are susceptible to antiviral drugs that inhibit the reverse transcriptase enzyme, e.g. zidovudine and lamivudine. An example of the first type is HIV, which is a retrovirus. Examples of the second type are the *Hepadnaviridae*, which includes Hepatitis B virus.

Effects on the Host Cell

The range of structural and biochemical effects that viruses have on the host cell is extensive. These are called *cytopathic effects*. Most virus infections eventually result in the death of the host cell. The causes of death include cell lysis, alterations to the cell's surface membrane and apoptosis. Often cell death is caused by cessation of its normal activities because of suppression by virus-specific proteins, not all of which are components of the virus particle.

Some viruses cause no apparent changes to the infected cell. Cells in which the virus is latent and inactive show few signs of infection and often function normally. This causes persistent infections and the virus is often dormant for many months or years. This is often the case with herpes viruses. Some viruses, such as Epstein-Barr virus, can cause cells to proliferate without causing malignancy, while others, such as papillomaviruses, are established causes of cancer.

Host Range

Viruses are by far the most abundant parasites on earth, and they have been found to infect all types of cellular life including animals, plants, bacteria and fungi. However, different types of viruses can infect only a limited range of hosts and many are species-specific. Some, such as smallpox virus for example, can infect only one species – in this case humans, and are said to have a narrow host range. Other viruses, such as rabies virus, can infect different species of mammals and are said to have a broad range. The viruses that infect plants are harmless to animals, and most viruses that infect other animals are harmless to humans. The host range of some bacteriophages is limited to a single strain of bacteria and they can be used to trace the source of outbreaks of infections by a method called phage typing.

Classification

Classification seeks to describe the diversity of viruses by naming and grouping them on the basis of similarities. In 1962, André Lwoff, Robert Horne, and Paul Tournier were the first to develop a means of virus classification, based on the Linnaean hierarchical system. This system bases classification on phylum, class, order, family, genus, and species. Viruses were grouped according to their shared properties (not those of their hosts) and the type of nucleic acid forming their genomes. Later the International Committee on Taxonomy of Viruses was formed. However, viruses are not classified on the basis of phylum or class, as their small genome size and high rate of mutation makes it difficult to determine their ancestry beyond Order. As such, the Baltimore Classification is used to supplement the more traditional hierarchy.

ICTV Classification

The International Committee on Taxonomy of Viruses (ICTV) developed the current classification system and wrote guidelines that put a greater weight on certain virus properties to maintain family uniformity. A unified taxonomy (a universal system for classifying viruses) has been established. The 7th lCTV Report formalised for the first time the concept of the virus species as the lowest taxon (group) in a branching hierarchy of viral taxa. However, at present only a small part of the total diversity of viruses has been studied, with analyses of samples from humans finding that about 20% of the virus sequences recovered have not been seen before, and samples from the environment, such as from seawater and ocean sediments, finding that the large majority of sequences are completely novel.

The general taxonomic structure is as follows:

- Order (-virales)
- Family (-viridae)

- Subfamily (-virinae)
- Genus (*-virus*)
- Species (*-virus*).

In the current (2010) ICTV taxonomy, six orders have been established, the Caudovirales, Herpesvirales, Mononegavirales, Nidovirales, Picornavirales and Tymovirales. A seventh order Ligamenvirales has also been proposed. The committee does not formally distinguish between subspecies, strains, and isolates. In total there are 6 orders, 87 families, 19 subfamilies, 348 genera, 2,885 species and about 3,000 types yet unclassified.

Baltimore Classification

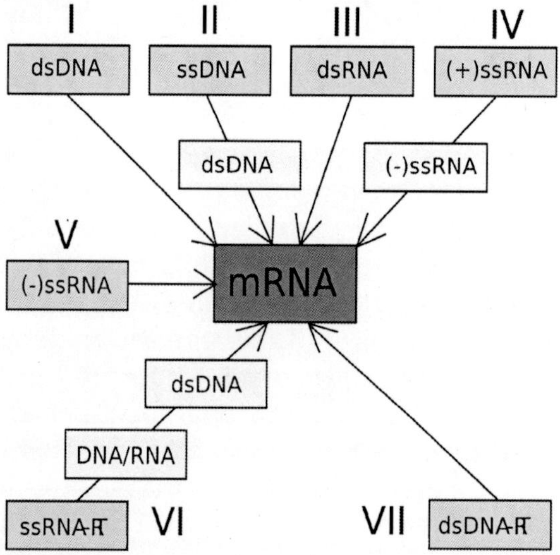

***Figure :** The Baltimore Classification of viruses is based on the method of viral mRNA synthesis.*

The Nobel Prize-winning biologist David Baltimore devised the Baltimore classification system. The ICTV classification system is used in conjunction with the Baltimore classification system in modern virus classification.

The Baltimore classification of viruses is based on the mechanism of mRNA production. Viruses must generate mRNAs from their genomes to produce proteins and replicate themselves, but different mechanisms are used to achieve this in each virus family. Viral genomes may be single-stranded (ss) or double-stranded (ds), RNA or DNA, and may or may not use reverse transcriptase (RT). Additionally, ssRNA viruses may be either sense (+) or antisense (“). This classification places viruses into seven groups:

- I: dsDNA viruses (e.g. Adenoviruses, Herpesviruses, Poxviruses)
- II: ssDNA viruses (+)sense DNA (e.g. Parvoviruses)

- III: dsRNA viruses (e.g. Reoviruses)
- IV: (+)ssRNA viruses (+)sense RNA (e.g. Picornaviruses, Togaviruses)
- V: (")ssRNA viruses (—)sense RNA (e.g. Orthomyxoviruses, Rhabdoviruses)
- VI: ssRNA-RT viruses (+)sense RNA with DNA intermediate in life-cycle (e.g. Retroviruses)
- VII: dsDNA-RT viruses (e.g. Hepadnaviruses).

Role in Human Disease

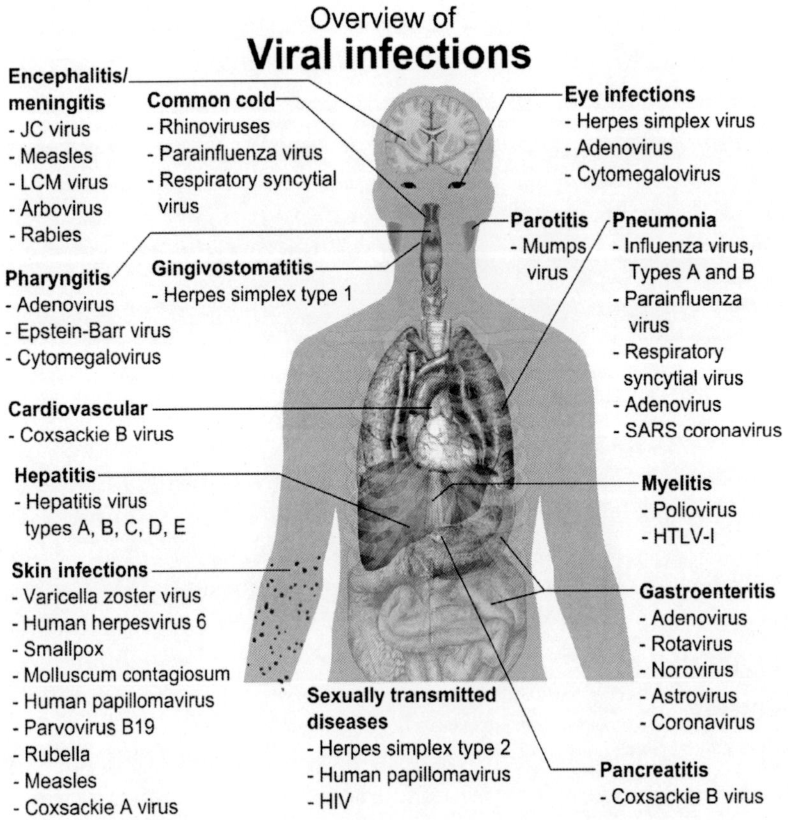

Figure : *Overview of the main types of viral infection and the most notable species involved.*

As an example of viral classification, the chicken pox virus, varicella zoster (VZV), belongs to the order Herpesvirales, family *Herpesviridae*, subfamily *Alphaherpesvirinae*, and genus *Varicellovirus*. VZV is in Group I of the Baltimore Classification because it is a dsDNA virus that does not use reverse transcriptase.

Examples of common human diseases caused by viruses include the common cold, influenza, chickenpox and cold sores. Many serious diseases such as ebola, AIDS, avian influenza and SARS are caused by viruses. The relative ability of viruses to cause disease is described in terms of virulence. Other diseases are under investigation as to whether they too have a virus as the causative agent, such as the possible connection between human herpes virus six (HHV6) and neurological diseases such as multiple sclerosis and chronic fatigue syndrome. There is controversy over whether the borna virus, previously thought to cause neurological diseases in horses, could be responsible for psychiatric illnesses in humans.

Viruses have different mechanisms by which they produce disease in an organism, which largely depends on the viral species. Mechanisms at the cellular level primarily include cell lysis, the breaking open and subsequent death of the cell. In multicellular organisms, if enough cells die, the whole organism will start to suffer the effects.

Although viruses cause disruption of healthy homeostasis, resulting in disease, they may exist relatively harmlessly within an organism. An example would include the ability of the herpes simplex virus, which causes cold sores, to remain in a dormant state within the human body.

This is called latency and is a characteristic of the herpes viruses including Epstein-Barr virus, which causes glandular fever, and varicella zoster virus, which causes chickenpox and shingles. Most people have been infected with at least one of these types of herpes virus. However, these latent viruses might sometimes be beneficial, as the presence of the virus can increase immunity against bacterial pathogens, such as *Yersinia pestis*. Some viruses can cause lifelong or chronic infections, where the viruses continue to replicate in the body despite the host's defence mechanisms. This is common in hepatitis B virus and hepatitis C virus infections. People chronically infected are known as carriers, as they serve as reservoirs of infectious virus. In populations with a high proportion of carriers, the disease is said to be endemic.

Epidemiology

Viral epidemiology is the branch of medical science that deals with the transmission and control of virus infections in humans. Transmission of viruses can be vertical, that is from mother to child, or horizontal, which means from person to person. Examples of vertical transmission include hepatitis B virus and HIV where the baby is born already infected with the virus. Another, more rare, example is the varicella zoster virus, which, although causing relatively mild infections in humans, can be fatal to the foetus and newborn baby.

Horizontal transmission is the most common mechanism of spread of viruses in populations. Transmission can occur when: body fluids are exchanged during sexual activity, e.g., HIV; blood is exchanged by contaminated transfusion or needle sharing, e.g., hepatitis C; a child is born to an infected mother, e.g., hepatitis B; exchange of saliva by mouth, e.g., Epstein-Barr virus; contaminated food or water is ingested, e.g., norovirus; aerosols containing virions are inhaled, e.g., influenza virus; and insect vectors such as mosquitoes penetrate the skin of a host, e.g., dengue. The rate or speed of transmission of viral infections depends on factors that include population density, the number of susceptible individuals, (i.e., those not immune), the quality of healthcare and the weather.

Epidemiology is used to break the chain of infection in populations during outbreaks of viral diseases. Control measures are used that are based on knowledge of how the virus is transmitted. It is important to find the source, or sources, of the outbreak and to identify the virus. Once the virus has been identified, the chain of transmission can sometimes be broken by vaccines. When vaccines are not available sanitation and disinfection can be effective.

Often infected people are isolated from the rest of the community and those that have been exposed to the virus placed in quarantine. To control the outbreak of foot and mouth disease in cattle in Britain in 2001, thousands of cattle were slaughtered.

Most viral infections of humans and other animals have incubation periods during which the infection causes no signs or symptoms. Incubation periods for viral diseases range from a few days to weeks but are known for most infections. Somewhat overlapping, but mainly following the incubation period, there is a period of communicability; a time when an infected individual or animal is contagious and can infect another person or animal. This too is known for many viral infections and knowledge the length of both periods is important in the control of outbreaks. When outbreaks cause an unusually high proportion of cases in a population, community or region they are called epidemics. If outbreaks spread worldwide they are called pandemics.

Epidemics and Pandemics

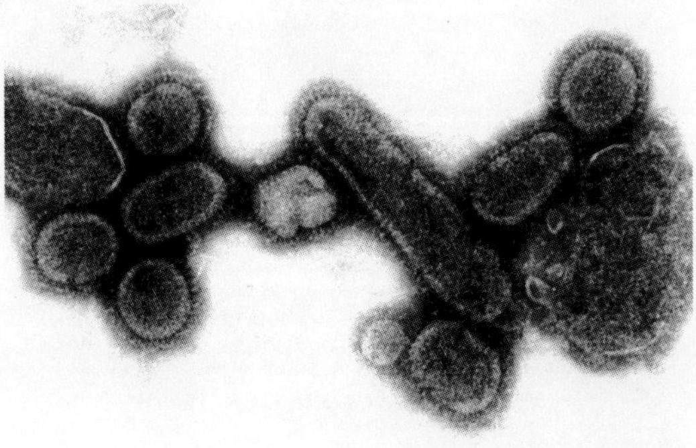

Figure : *Transmission electron microscope image of a recreated 1918 influenza virus*

Native American populations were devastated by contagious diseases, in particular, smallpox, brought to the Americas by European colonists. It is unclear how many Native Americans were killed by foreign diseases after the arrival of Columbus in the Americas, but the numbers have been estimated to be close to 70% of the indigenous population. The damage done by this disease significantly aided European attempts to displace and conquer the native population.

A pandemic is a worldwide epidemic. The 1918 flu pandemic, commonly referred to as the Spanish flu, was a category 5 influenza pandemic caused by an unusually severe and deadly influenza A virus. The victims were often healthy young adults, in contrast to most influenza outbreaks, which predominantly affect juvenile, elderly, or otherwise-weakened patients.

The Spanish flu pandemic lasted from 1918 to 1919. Older estimates say it killed 40–50 million people, while more recent research suggests that it may have killed as many as 100 million people, or 5% of the world's population in 1918.

Most researchers believe that HIV originated in sub-Saharan Africa during the 20th century; it is now a pandemic, with an estimated 38.6 million people now living with the disease worldwide. The Joint United Nations Programme on HIV/AIDS (UNAIDS) and the World Health Organisation (WHO) estimate that AIDS has killed more than 25 million people since it was first recognised on June 5, 1981, making it one of the most destructive epidemics in recorded history. In 2007 there were 2.7 million new HIV infections and 2 million HIV-related deaths.

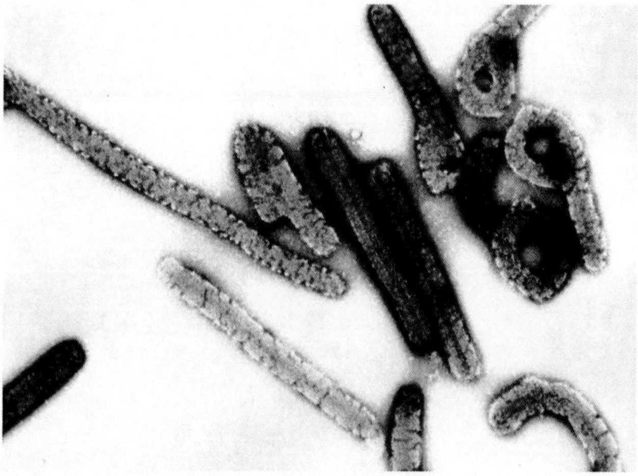

***Figure:** Marburg virus*

Several highly lethal viral pathogens are members of the *Filoviridae*. Filoviruses are filament-like viruses that cause viral hemorrhagic fever, and include the ebola and marburg viruses. The Marburg virus attracted widespread press attention in April 2005 for an outbreak in Angola. Beginning in October 2004 and continuing into 2005, the outbreak was the world's worst epidemic of any kind of viral hemorrhagic fever.

Cancer

Viruses are an established cause of cancer in humans and other species. Viral cancers occur only in a minority of infected persons (or animals). Cancer viruses come from a range of virus families, including both RNA and DNA viruses, and so there is no single type of "oncovirus" (an obsolete term originally used for acutely transforming retroviruses).

The development of cancer is determined by a variety of factors such as host immunity and mutations in the host. Viruses accepted to cause human cancers include some genotypes of human papillomavirus, hepatitis B virus, hepatitis C virus, Epstein-Barr virus, Kaposi's sarcoma-associated herpesvirus and human T-lymphotropic virus. The most recently discovered human cancer virus is a polyomavirus (Merkel cell polyomavirus) that causes most cases of a rare form of skin cancer called Merkel cell carcinoma.

Hepatitis viruses can develop into a chronic viral infection that leads to liver cancer. Infection by human T-lymphotropic virus can lcad to tropical spastic paraparesis and adult T-cell leukemia. Human papillomaviruses are an established cause of cancers of cervix, skin, anus, and penis. Within the *Herpesviridae*, Kaposi's sarcoma-associated herpesvirus causes Kaposi's sarcoma and body cavity lymphoma, and Epstein–Barr virus causes Burkitt's lymphoma, Hodgkin's lymphoma, B lymphoproliferative disorder, and nasopharyngeal carcinoma. Merkel cell polyomavirus closely related to SV40 and mouse polyomaviruses that have been used as animal models for cancer viruses for over 50 years.

Host Defence Mechanisms

The body's first line of defence against viruses is the innate immune system. This comprises cells and other mechanisms that defend the host from infection in a non-specific manner. This means that the cells of the innate system recognise, and respond to, pathogens in a generic way, but, unlike the adaptive immune system, it does not confer long-lasting or protective immunity to the host.

RNA interference is an important innate defence against viruses. Many viruses have a replication strategy that involves double-stranded RNA (dsRNA). When such a virus infects a cell, it releases its RNA molecule or molecules, which immediately bind to a protein complex called dicer that cuts the RNA into smaller pieces. A biochemical pathway called the RISC complex is activated, which degrades the viral mRNA and the cell survives the infection. Rotaviruses avoid this mechanism by not uncoating fully inside the cell and by releasing newly produced mRNA through pores in the particle's inner capsid. The genomic dsRNA remains protected inside the core of the virion.

When the adaptive immune system of a vertebrate encounters a virus, it produces specific antibodies that bind to the virus and render it non-infectious. This is called humoral immunity. Two types of antibodies are important. The first, called IgM, is highly effective at neutralizing viruses but is produced by the cells of the immune system only for a few weeks.

The second, called IgG, is produced indefinitely. The presence of IgM in the blood of the host is used to test for acute infection, whereas IgG indicates an infection sometime in the past. IgG antibody is measured when tests for immunity are carried out. A second defence of vertebrates against viruses is called cell-mediated immunity and involves immune cells known as T cells. The body's cells constantly display short fragments of their proteins on the cell's surface, and, if a T cell recognises a suspicious viral fragment there, the host cell is destroyed by *killer T* cells and the virus-specific T-cells proliferate.

Cells such as the macrophage are specialists at this antigen presentation. The production of interferon is an important host defence mechanism. This is a hormone produced by the body when viruses are present. Its role in immunity is complex; it eventually stops the viruses from reproducing by killing the infected cell and its close neighbours.

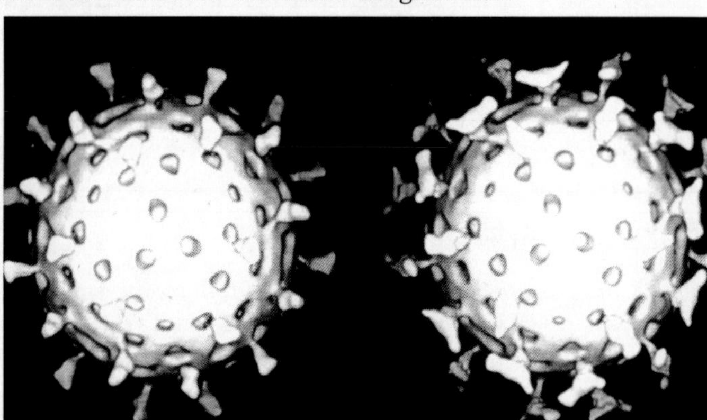

Figure: *Two rotaviruses: the one on the right is coated with antibodies that stop its attaching to cells and infecting them*

Not all virus infections produce a protective immune response in this way. HIV evades the immune system by constantly changing the amino acid sequence of the proteins on the surface of the virion. These persistent viruses evade immune control by sequestration, blockade of antigen presentation, cytokine resistance, evasion of natural killer cell activities, escape from apoptosis, and antigenic shift. Other viruses, called *neurotropic viruses*, are disseminated by neural spread where the immune system may be unable to reach them.

Prevention and Treatment

Because viruses use vital metabolic pathways within host cells to replicate, they are difficult to eliminate without using drugs that cause toxic effects to host cells in general. The most effective medical approaches to viral diseases are vaccinations to provide immunity to infection, and antiviral drugs that selectively interfere with viral replication.

Vaccines

Vaccination is a cheap and effective way of preventing infections by viruses. Vaccines were used to prevent viral infections long before the discovery of the actual viruses. Their use has resulted in a dramatic decline in morbidity (illness) and mortality (death) associated with viral infections such as polio, measles, mumps and rubella.

Antiviral Drugs

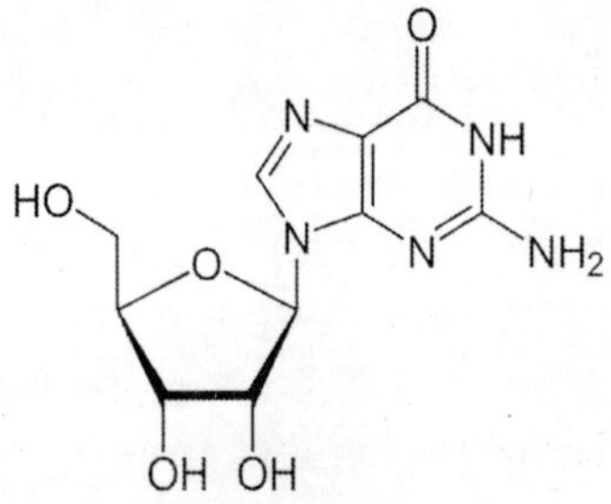

Figure: *Guanosine*

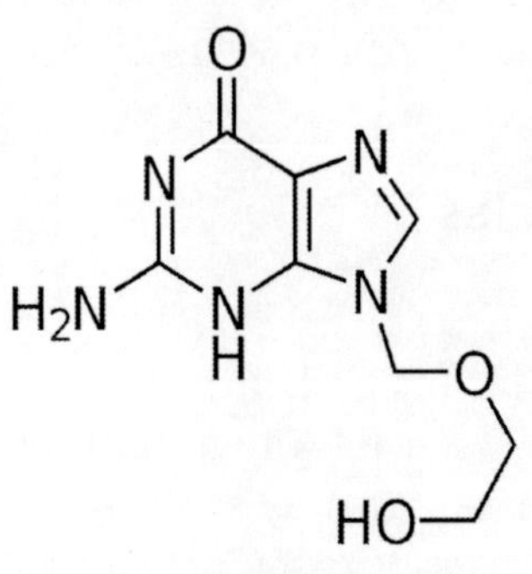

Figure: *The guanosine analogue Aciclovir*

Smallpox infections have been eradicated. Vaccines are available to prevent over thirteen viral infections of humans, and more are used to prevent viral infections of animals. Vaccines can consist of live-attenuated or killed viruses, or viral proteins (antigens). Live vaccines contain weakened forms of the virus, which do not cause the disease but, nonetheless, confer immunity. Such viruses are called attenuated. Live vaccines can be dangerous when given to people with a weak immunity, (who are described as immunocompromised), because in these people, the weakened virus can cause the original disease. Biotechnology and genetic engineering techniques are used to produce subunit vaccines. These vaccines use only the capsid proteins of the virus.

Hepatitis B vaccine is an example of this type of vaccine. Subunit vaccines are safe for immunocompromised patients because they cannot cause the disease. The yellow fever virus vaccine, a live-attenuated strain called 17D, is probably the safest and most effective vaccine ever generated.

Antiviral drugs are often nucleoside analogues, (fake DNA building-blocks), which viruses mistakenly incorporate into their genomes during replication. The life-cycle of the virus is then halted because the newly synthesised DNA is inactive. This is because these analogues lack the hydroxyl groups, which, along with phosphorus atoms, link together to form the strong "backbone" of the DNA molecule. This is called DNA chain termination. Examples of nucleoside analogues are aciclovir for Herpes simplex virus infections and lamivudine for HIV and Hepatitis B virus infections. Aciclovir is one of the oldest and most frequently prescribed antiviral drugs. Other antiviral drugs in use target different stages of the viral life cycle. HIV is dependent on a proteolytic enzyme called the HIV-1 protease for it to become fully infectious. There is a large class of drugs called protease inhibitors that inactivate this enzyme.

Hepatitis C is caused by an RNA virus. In 80% of people infected, the disease is chronic, and without treatment, they are infected for the remainder of their lives. However, there is now an effective treatment that uses the nucleoside analogue drug ribavirin combined with interferon. The treatment of chronic carriers of the hepatitis B virus by using a similar strategy using lamivudine has been developed.

Infection in Other Species

Plants

Viruses infect all cellular life and, although viruses occur universally, each cellular species has its own specific range that often infect only that species. Some viruses, called satellites, can only replicate within cells that have already been infected by another virus. Viruses are important pathogens of livestock. Diseases such as Foot and Mouth Disease and bluetongue are caused by viruses. Companion animals such as cats, dogs, and horses, if not vaccinated, are susceptible to serious viral infections. Canine parvovirus is caused by a small DNA virus and infections are often fatal in pups. Like all invertebrates, the honey bee is susceptible to many viral infections. Fortunately, most viruses co-exist harmlessly in their host and cause no signs or symptoms of disease. There are many types of plant virus, but often they cause only a loss of yield, and it is not economically viable to try to control them. Plant viruses are often spread from plant to plant by organisms, known as *vectors*.

Figure: *Peppers infected by mild mottle virus*

These are normally insects, but some fungi, nematode worms, and single-celled organisms have been shown to be vectors. When control of plant virus infections is considered economical, for perennial fruits, for example, efforts are concentrated on killing the vectors and removing alternate hosts such as weeds. Plant viruses are harmless to humans and other animals because they can reproduce only in living plant cells.

Plants have elaborate and effective defence mechanisms against viruses. One of the most effective is the presence of so-called resistance (R) genes. Each R gene confers resistance to a particular virus by triggering localised areas of cell death around the infected cell, which can often be seen with the unaided eye as large spots. This stops the infection from spreading. RNA interference is also an effective defence in plants. When they are infected, plants often produce natural disinfectants that kill viruses, such as salicylic acid, nitric oxide, and reactive oxygen molecules.

Plant virus particles or virus-like particles (VLPs) have applications in both biotechnology and nanotechnology. The capsids of most plant viruses are simple and robust structures and can be produced in large quantities either by the infection of plants or by expression in a variety of heterologous systems. Plant virus particles can be modified genetically and chemically to encapsulate foreign material and can be incorporated into supramolecular structures for use in biotechnology.

Bacteria

Bacteriophages are a common and diverse group of viruses and are the most abundant form of biological entity in aquatic environments – there are up to ten times more of these viruses in the oceans than there are bacteria, reaching levels of 250,000,000 bacteriophages per millilitre of seawater. These viruses infect specific bacteria by binding to surface receptor molecules and then entering the cell.

Within a short amount of time, in some cases just minutes, bacterial polymerase starts translating viral mRNA into protein. These proteins go on to become either new virions within the cell, helper proteins, which help assembly of new virions, or proteins involved in cell lysis. Viral enzymes aid in the breakdown of the cell membrane, and, in the case of the T4 phage, in just over twenty minutes after injection over three hundred phages could be released.

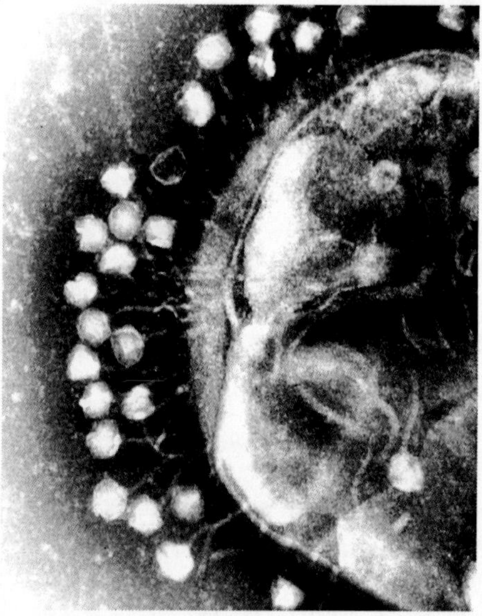

Figure : *Transmission electron micrograph of multiple bacteriophages attached to a bacterial cell wall*

The major way bacteria defend themselves from bacteriophages is by producing enzymes that destroy foreign DNA. These enzymes, called restriction endonucleases, cut up the viral DNA that bacteriophages inject into bacterial cells. Bacteria also contain a system that uses CRISPR sequences to retain fragments of the genomes of viruses that the bacteria have come into contact with in the past, which allows them to block the virus's replication through a form of RNA interference. This genetic system provides bacteria with acquired immunity to infection.

Archaea

Some viruses replicate within archaea: these are double-stranded DNA viruses with unusual and sometimes unique shapes.

These viruses have been studied in most detail in the thermophilic archaea, particularly the orders Sulfolobales and Thermoproteales. Defences against these viruses may involve RNA interference from repetitive DNA sequences within archaean genomes that are related to the genes of the viruses.

Role in Aquatic Ecosystems

Viruses are the most abundant biological entity in aquatic environments: a teaspoon of seawater contains about one million of them. They are essential to the regulation of saltwater and freshwater ecosystems.

Most of these viruses are bacteriophages, which are harmless to plants and animals. They infect and destroy the bacteria in aquatic microbial communities, comprising the most important mechanism of recycling carbon in the marine environment. The organic molecules released from the bacterial cells by the viruses stimulates fresh bacterial and algal growth.

Microorganisms constitute more than 90% of the biomass in the sea. It is estimated that viruses kill approximately 20% of this biomass each day and that there are 15 times as many viruses in the oceans as there are bacteria and archaea. Viruses are the main agents responsible for the rapid destruction of harmful algal blooms, which often kill other marine life.

The number of viruses in the oceans decreases further offshore and deeper into the water, where there are fewer host organisms. The effects of marine viruses are far-reaching; by increasing the amount of photosynthesis in the oceans, viruses are indirectly responsible for reducing the amount of carbon dioxide in the atmosphere by approximately 3 gigatonnes of carbon per year.

Like any organism, marine mammals are susceptible to viral infections. In 1988 and 2002, thousands of harbour seals were killed in Europe by phocine distemper virus. Many other viruses, including caliciviruses, herpesviruses, adenoviruses and parvoviruses, circulate in marine mammal populations.

Role in Evolution

Viruses are an important natural means of transferring genes between different species, which increases genetic diversity and drives evolution. It is thought that viruses played a central role in the early evolution, before the diversification of bacteria, archaea and eukaryotes and at the time of the last universal common ancestor of life on Earth. Viruses are still one of the largest reservoirs of unexplored genetic diversity on Earth.

Applications

Life Sciences and Medicine

Viruses are important to the study of molecular and cellular biology as they provide simple systems that can be used to manipulate and investigate the functions of cells.

The study and use of viruses have provided valuable information about aspects of cell biology. For example, viruses have been useful in the study of genetics and helped our understanding of the basic mechanisms of molecular genetics, such as DNA replication, transcription, RNA processing, translation, protein transport, and immunology.

Geneticists often use viruses as vectors to introduce genes into cells that they are studying. This is useful for making the cell produce a foreign substance, or to study the effect of introducing a new gene into the genome. In similar fashion, virotherapy uses viruses as vectors to treat various diseases, as they can specifically target cells and DNA. It shows promising use in the treatment of cancer and in gene therapy. Eastern European scientists have used phage therapy as an alternative to antibiotics for some time, and interest in this approach is increasing, because of the high level of antibiotic resistance now found in some pathogenic bacteria.

Expression of heterologous proteins by viruses is the basis of several manufacturing processes that are currently being used for the production of various proteins such as vaccine antigens and antibodies. Industrial processes have been recently developed using viral vectors and a number of pharmaceutical proteins are currently in preclinical and clinical trials.

Materials Science and Nanotechnology

Current trends in nanotechnology promise to make much more versatile use of viruses. From the viewpoint of a materials scientist, viruses can be regarded as organic nanoparticles. Their surface carries specific tools designed to cross the barriers of their host cells. The size and shape of viruses, and the number and nature of the functional groups on their surface, is precisely defined. As such, viruses are commonly used in materials science as scaffolds for covalently linked surface modifications. A particular quality of viruses is that they can be tailored by directed evolution. The powerful techniques developed by life sciences are becoming the basis of engineering approaches towards nanomaterials, opening a wide range of applications far beyond biology and medicine.

Because of their size, shape, and well-defined chemical structures, viruses have been used as templates for organising materials on the nanoscale. Recent examples include work at the Naval Research Laboratory in Washington, DC, using Cowpea Mosaic Virus (CPMV) particles to amplify signals in DNA microarray based sensors. In this application, the virus particles separate the fluorescent dyes used for signalling to prevent the formation of non-fluorescent dimers that act as quenchers. Another example is the use of CPMV as a nanoscale breadboard for molecular electronics.

Synthetic Viruses

Many viruses can be synthesized de novo ("from scratch") and the first synthetic virus was created in 2002. Although somewhat of a misconception, it is not the actual virus that is synthesized, but rather its DNA genome (in case of a DNA virus), or a cDNA copy of its genome (in case of RNA viruses). For many virus families the naked synthetic DNA or RNA (once enzymatically converted back from the synthetic cDNA) is infectious when introduced into a cell. That is, they contain all the necessary information to produce new viruses.

This technology is now being used to investigate novel vaccine strategies. The ability to synthesize viruses has far-reaching consequences, since viruses can no longer be regarded as extinct, as long as the information of their genome sequence is known and permissive cells are available. Currently, the full-length genome sequences of 2408 different viruses (including smallpox) are publicly available at an online database, maintained by the National Institute of Health.

Weapons

The ability of viruses to cause devastating epidemics in human societies has led to the concern that viruses could be weaponised for biological warfare. Further concern was raised by the successful recreation of the infamous 1918 influenza virus in a laboratory. The smallpox virus devastated numerous societies throughout history before its eradication.

There are officially only two centres in the world that keep stocks of smallpox virus – the Russian Vector laboratory, and the United States Centres for Disease Control. But fears that it may be used as a weapon are not totally unfounded; the vaccine for smallpox has sometimes severe side-effects – during the last years before the eradication of smallpox disease more people became seriously ill as a result of vaccination than did people from smallpox – and smallpox vaccination is no longer universally practiced. Thus, much of the modern human population has almost no established resistance to smallpox.

A viral Antigen is an antigen with multiple antigenicities that is protein in nature, strain-specific, and closely associated with the virus particle. A viral antigen is a protein encoded by the viral genome. A viral protein is an antigen specified by the viral genome that can be detected by a specific immunological response. Viruses are infectious pathogens that cause serious diseases & major threats for global public health, such as influenza, hepatitis, & AIDS. Virus is a sub-micrometer particle that has DNA or RNA packed in a shell called capsid.

Viral antigens protrude from the capsid and often fulfill important function in docking to the host cell, fusion, and injection of viral DNA/RNA. Antibody-based immune responses form a first layer of protection of the host from viral infection; however, in many cases a vigorous cellular immune response mediated by T-cells and NK-cells is required for effective viral clearance. When cellular immunity is unable to clear the virus, the infection can become chronic, and serum antibodies to the viral pathogen are used as first indicator for the diagnosis of the disease.

ELISAs provide a valuable tool in the detection and diagnosis of virus infection. The ability to produce recombinant viral proteins will ensure that future ELISAs are safe, specific and rapid. Even when a virus cannot be cultured, provided gene sequence is available, it is possible to rapidly respond to emerging viruses and new viral strains of existing pathogens.

Recombinant viral antigens contain part of viral sequence meaning that the recombinant antigen contains a region which can be recognised by different antibodies produced by different individuals. This reduces the risk of false negatives which can occur with synthetic peptides, which contain only a small portion of the entire protein. If an individual infected with a viral antigen makes antibodies to a part of the protein not included in the synthetic peptides, a false negative results.

Recombinant viral protein usually contains a fusion protein/partner which produces superior attachment to assay surfaces such as wells. For this reason, smaller amounts of recombinant protein will produce the same results as larger amounts of unfused protein. The choice of fusion partner prevents false positives, allowing superior adhesion without incorrect results.

Recombinant Viral proteins are expressed in bacteria, yeast, mammalian cells, and viruses. E. Coli cells were first to be used for this purpose but the expressed proteins were not glycosylated, which was a major drawback since many of the immunogenic proteins of viruses such as the envelope glycoproteins, were glycosylated. Nevertheless, in many instances, it was demonstrated that the non-glycosylated protein backbone was just as immunogenic. The obvious advantage of recombinant viral antigens is that they are available in unlimited quantities and the production and quality control processes is simple.

Advantages of defined using recombinant viral antigens:

1. Production and quality control is simple.
2. No nucleic acids or other viral or external proteins, therefore less toxic.

3. Safer in cases where viruses are oncogenic or establish a persistent infection.
4. Feasible even if virus cannot be cultivated.

Disadvantages:

1. May be less immunogenic than conventional inactivated whole-virus vaccines.
2. Requires adjuvant .
3. Fails to elicit CMI.

Facts about Viral Antigens:

1. A Viral Protein Mimics its Way into cells.
2. Viral Protein Helps Infected T Cells Stick To Uninfected Cells.
3. The Viral Protein A238L Inhibits Cyclooxygenase-2 Expression through a Nuclear Factor of Activated T Cell-dependent Trans-activation Pathway.
4. Viral Protein is an effective preventative against ear infection.
5. HIV-1 Viral Protein R Induces Apoptosis via a Direct Effect on the Mitochondrial Permeability Transition Pore.
6. The Level of Viral Antigen Presented by Hepatocytes Influences CD8 T-Cell Function.
7. Antigen-presenting cells from calves persistently infected with bovine viral diarrhea virus, a member of the Flaviviridae, are not compromised in their ability to present viral antigen.
8. There is a difference in the distribution and spread of a viral antigen, development of lesions and correlation between presence of viral antigen and lesions.
9. The absence of viral antigens on the surface of equine herpesvirus-1-infected peripheral blood mononuclear cells is a strategy to avoid complement-mediated lysis.
10. Viral Protein Influences Key Cell-signalling Pathway.
11. A viral protein produced by cancer-causing virus influences a key signaling pathway in the immune cells that the virus infects. This stimulates the cells to divide, helping the virus spread through the body.
12 Protection by recombinant viral proteins against a respiratory virulent avian metapneumovirus has been achieved.
13. Viral O-acetylesterases are found in influenza C viruses and Corona-viruses. Viral O-acetylesterases remove cellular receptors from the

surface of target cells which destroys the receptor. Recombinant viral O-acetylesterases derived from Sf9 insect cells as chimeric proteins fused to eGFP specifically hydrolyze 9-O-acetylated sialic acids, while that of sialodacryoadenitis virus, a rat coronavirus related to mouse hepatitis virus, is specific for 4-O-acetylated sialic acid. The recombinant esterases were shown to specifically de-O-acetylate sialic acids on glycoconjugates. The recombinant viral proteins can be used to unambiguously identify O-acetylated acids.

An antigen is a foreign molecule that, when introduced into the body, triggers the production of an antibody by the immune system. The immune system will then kill or neutralize the antigen that is recognised as a foreign and potentially harmful invader. These invaders can be molecules such as pollen or cells such as bacteria.

The term originally came from antibody generator and was a molecule that binds specifically to an antibody, but the term now also refers to any molecule or molecular fragment that can be bound by a major histocompatibility complex (MHC) and presented to a T-cell receptor. "Self" antigens are usually tolerated by the immune system; whereas "Non-self" antigens are identified as invaders and attacked by the immune system.

An immunogen is a specific type of antigen. An immunogen is a substance that is able to provoke an adaptive immune response if injected on its own. An immunogen is able to induce an immune response, whereas an antigen is able to combine with the products of an immune response once they are made. The overlapping concepts of immunogenicity and antigenicity are, therefore, subtly different. According to a current textbook:

Immunogenicity is the ability to induce a humoral and/or cell-mediated immune response.

Antigenicity is the ability to combine specifically with the final products of the immune response (i.e. secreted antibodies and/or surface receptors on T-cells). Although all molecules that have the property of immunogenicity also have the property of antigenicity, the reverse is not true." At the molecular level, an antigen is characterised by its ability to be "bound" at the antigen-binding site of an antibody. Note also that antibodies tend to discriminate between the specific molecular structures presented on the surface of the antigen. Antigens are usually proteins or polysaccharides. This includes parts (coats, capsules, cell walls, flagella, fimbrae, and toxins) of bacteria, viruses, and other microorganisms. Lipids and nucleic acids are antigenic only when combined with proteins and polysaccharides. Non-microbial exogenous (non-self) antigens can include pollen, egg white, and proteins from transplanted tissues and organs or on the surface of transfused blood cells. Vaccines are examples of immunogenic antigens intentionally administered to induce acquired immunity in the recipient. Cells present their immunogenic-antigens to the immune system via a histocompatibility molecule. Depending on the antigen presented and the type of the histocompatibility molecule, several types of immune cells can become activated.

Epitope

An epitope, also known as *antigenic determinant*, is the part of an antigen that is recognised by the immune system, specifically by antibodies, B cells, or T cells. The part of an antibody that recognises the epitope is called a paratope. Although epitopes are usually thought to be derived from non-self proteins, sequences derived from the host that can be recognised are also classified as epitopes.

The epitopes of protein antigens are divided into two categories, conformational epitopes and linear epitopes, based on their structure and interaction with the paratope. A conformational epitope is composed of discontinuous sections of the antigen's amino acid sequence. These epitopes interact with the paratope based on the 3-D surface features and shape or tertiary structure of the antigen. Most epitopes are conformational. By contrast, linear epitopes interact with the paratope based on their primary structure. A linear epitope is formed by a continuous sequence of amino acids from the antigen.

The distinct molecular surface features of an antigen capable of being bound by an antibody (a.k.a. *antigenic determinant*). Antigenic molecules, normally being "large" biological polymers, usually present several surface features that can act as points of interaction for specific antibodies. Any such distinct molecular feature constitutes an epitope. Therefore, most antigens have the potential to be bound by several distinct antibodies, each of which is specific to a particular epitope. Using the "lock and key" metaphor, the antigen itself can be seen as a string of keys - any epitope being a "key" - each of which can match a different lock. Different antibody idiotypes, each having distinctly formed complementarity determining regions, correspond to the various "locks" that can match "the keys" (epitopes) presented on the antigen molecule.

Function

T cell Epitopes

T cell epitopes are presented on the surface of an antigen-presenting cell, where they are bound to MHC molecules. T cell epitopes presented by MHC class I molecules are typically peptides between 8 and 11 amino acids in length, whereas MHC class II molecules present longer peptides, and non-classical MHC molecules also present non-peptidic epitopes such as glycolipids.

Cross-reactivity

Epitopes are sometimes cross-reactive. This property is exploited by the immune system in regulation by anti-idiotypic antibodies (originally proposed by Nobel laureate Niels Kaj Jerne). If an antibody binds to an antigen's epitope, the paratope could become the epitope for another antibody that will then bind to it. If this second antibody is of IgM class, its binding can upregulate the immune response; if the second antibody is of IgG class, its binding can downregulate the immune response.

Epitope Mapping

Epitopes can be mapped using protein microarrays, and with the ELISPOT or ELISA techniques.

Intensive research is currently taking place to design reliable tools that will predict epitopes on proteins.

Epitope mapping is the process of identifying the binding sites, or 'epitopes', of antibodies on their target antigens (which are proteins).

Identification and characterisation of the binding sites of antibodies can aid in the discovery and development of new therapeutics, vaccines, and diagnostics.

Epitopes (the binding sites on the protein) can be divided into linear and conformational. *Linear epitopes* are formed by a continuous sequence of amino acids in a protein, while *conformational epitopes* are composed of amino acids that are discontinuous in the protein sequence but are brought together upon three-dimensional protein folding. The vast majority of antigen-antibody interactions have conformational epitopes.

Methods for Epitope Mapping

Epitope mapping of complex target antigens, such as integral membrane proteins or multi-subunit proteins, is often challenging because of the difficulty in expressing and purifying these types of antigens.

There are several methods available for mapping antibody epitopes on target antigens :

- The gold standard approach is X-ray co-crystallography, which allows direct visualisation of the interaction between the antigen and antibody. However, this approach is technically challenging, requires large amounts of purified protein, and can be time-consuming and expensive.
- Array-based oligo-peptide scanning (sometimes called pepscan analysis): This technique uses a library of oligo-peptide sequences from overlapping and non-overlapping segments of a target protein and tests for their ability to bind the antibody of interest. This method is fast and relatively

inexpensive, and specifically suited to profile epitopes for large number of candidate antibodies against a defined target. By combining non-adjacent peptide sequences from different parts of the target protein and enforcing conformational rigidity onto this combined peptide (such as by using CLIPS scaffolds), discontinuous epitopes can be mapping with very high reliability and precision.

- Site-directed mutagenesis : Using this approach, systematic mutations of amino acids are introduced into a protein sequence followed by measurement of antibody binding in order to identify amino acids that comprise an epitope. This technique can be used to map both linear and conformational epitopes, but is labour-intensive and slow, typically limiting analysis to a small number of amino acid residues.
- Mutagenesis Mapping. This approach utilises a comprehensive mutation library, with each clone containing a unique amino acid mutation and the entire library covering every amino acid in the target protein. Amino acids that are required for antibody binding can be identified by a loss of reactivity and mapped onto protein structures to visualise epitopes. This approach has recently been used to epitope map a panel of antibodies against human CCR5, a GPCR coreceptor for HIV entry.
- Other methods, such as phage display, and limited proteolysis, provide high throughput but lack reliability, especially for conformational epitopes.

Epitope Tags

Epitopes are often used in proteomics and the study of other gene products. Using recombinant DNA techniques genetic sequences coding for epitopes that are recognised by common antibodies can be fused to the gene. Following synthesis, the resulting epitope tag allows the antibody to find the protein or other gene product enabling lab techniques for localisation, purification, and further molecular characterisation. Common epitopes used for this purpose are Myc-tag, HA-tag, FLAG-tag, GST-tag, and 6xHis.

- Allergen - A substance capable of causing an allergic reaction. The (detrimental) reaction may result after exposure via ingestion, inhalation, injection, or contact with skin.
- Superantigen - A class of antigens that cause non-specific activation of T-cells, resulting in polyclonal T cell activation and massive cytokine release.

Superantigens (SAgs) are a class of antigens which cause non-specific activation of T-cells resulting in oligoclonal T cell activation and massive cytokine release. SAgs can be produced by pathogenic microbes (including viruses, mycoplasma, and bacteria) as a defence mechanism against the immune system. Compared to a normal antigen-induced T-cell response where .001-.0001% of the body's T-cells are activated, these SAgs are capable of activating up to 20% of the body's T-cells. Furthermore, Anti-CD3 and Anti-CD28 Antibodies (CD28-SuperMAB) have also shown to be highly potent superantigens (and can activate up to 100% of T cells).

The large number of activated T-cells generates a massive immune response which is not specific to any particular epitope on the SAg thus undermining one of the fundamental strengths of the adaptive immune system, that is, its ability to target antigens with high specificity. More importantly, the large number of activated T-cells secrete large amounts of cytokines (the most important of which is TNF-alpha).

TNF-alpha is particularly important as a part of the body's inflammatory response, and in normal circumstances (where it is released locally in low levels) helps the immune system defeat pathogens. However when it is systemically released in the blood and in high levels (due to mass T-cell activation resulting from the SAg binding), it can cause severe and life-threatening symptoms, including shock and multiple organ failure.

Structure

SAgs are produced intracellularly by bacteria and are released upon infection as extracellular mature toxins.

The sequences of these toxins are relatively conserved among the different subgroups. More important than sequence homology, the 3D structure is very similar among different SAgs resulting in similar functional effects among different groups.

Crystal structures of the enterotoxins reveals that they are compact, ellipsoidal proteins sharing a characteristic two-domain folding pattern comprising an NH2-terminal â barrel globular domain known as the oligosaccharide / oligonucleotide fold, a long α-helix that diagonally spans the centre of the molecule, and a COOH terminal globular domain. The domains have binding regions for the Major Histocompatibility Complex Class II (MHC Class II) and the T cell receptor (TCR), respectively.

Binding

Superantigens bind first to the MHC Class II and then coordinate to the variable alpha or beta chain of T-cell Receptors (TCR)

MHC Class II

SAgs show preference for the HLA-DQ form of the molecule. Binding to the α-chain puts the SAg in the appropriate position to coordinate to the TCR.

Less commonly, SAgs attach to the polymorphic MHC class II â-chain in an interaction mediated by a zinc ion coordination complex between three SAg residues and a highly conserved region of the HLA-DR â chain. The use of a zinc ion in binding leads to a higher affinity interaction. Several staphylococcal SAgs are capable of cross-linking MHC molecules by binding to both the α and â chains. This mechanism stimulates cytokine expression and release in antigen presenting cells as well as inducing the production of costimulatory molecules that allow the cell to bind to and activate T cells more effectively.

T-cell Receptor

T-cell binding region of the SAg interacts with the Variable region on the Beta chain of the T-cell Receptor. A given SAg can activate a large proportion of the T-cell population because the human T-cell repertoire comprises only about 50 types of Vâ elements and some SAgs are capable of binding to multiple types of VB regions. This interaction varies slightly among the different groups of SAgs. Variability among different people in the types of T-cell regions that are prevalent explains why some people respond more strongly to certain SAgs.

Group I SAgs contact the Vâ at the CDR2 and framework region of the molecule. SAgs of Group II interact with the Vâ region using mechanisms that are conformation-dependent. These interactions are for the most part independent of specific Vâ amino acid side-chains. Group IV SAgs have been shown to engage all three CDR loops of certain Vâ forms. The interaction takes place in a cleft between the small and large domains of the SAg and allows the SAg to act as a wedge between the TCR and MHC. This displaces the antigenic peptide away from the TCR and circumvents the normal mechanism for T-cell activation.

The biological strength of the SAg (its ability to stimulate) is determined by its affinity for the TCR. SAgs with the highest affinity for the TCR elicit the strongest response. SPMEZ-2 is the most potent SAg discovered to date.

T-cell Signaling

The SAg cross-links the MHC and the TCR inducing a signaling pathway that results in the proliferation of the cell and production of cytokines. Low levels of Zap-70 have been found in T-cells activated by SAgs, indicating that the normal signaling pathway of T-cell activation is impaired.

It is hypothesized that Fyn rather than Lck is activated by a tyrosine kinase, leading to the adaptive induction of anergy.

Both the protein kinase C pathway and the protein tyrosine kinase pathways are activated, resulting in upregulating production of proinflammatory cytokines.

This alternative signaling pathway impairs the calcium/calcineurin and Ras/MAPkinase pathways slightly, but allows for a focused inflammatory response.

Direct Effects

SAg stimulation of antigen presenting cells and T-cells elicits a response that is mainly inflammatory, focused on the action of Th1 T-helper cells. Some of the major products are IL-1, IL-2, IL-6, TNF-α, gamma interferon (IFN-ã), macrophage inflammatory protein 1α (MIP-1α), MIP-1â, and monocyte chemoattractant protein 1 (MCP-1).

This excessive uncoordinated release of cytokines, (especially TNF-α), overloads the body and results in rashes, fever, and can lead to multi-organ failure, coma and death.

Deletion or anergy of activated T-cells follows infection. This results from production of IL-10 from prolonged exposure to the toxin. IL-10 downregulates production of IL-2, MHC Class II, and costimulatory molecules on the surface of APCs. These effects produce memory cells that are unresponsive to antigen stimulation.

One mechanism by which this is possible involves cytokine-mediated suppression of T-cells. MHC crosslinking also activates a signaling pathway that suppresses hematopoiesis and upregulates Fas-mediated apoptosis. IFN-α is another product of prolonged SAg exposure. This cytokine is closely linked with induction of autoimmunity, and the autoimmune disease Kawasaki Disease is known to be caused by SAg infection.

SAg activation in T-cells leads to production of CD40 ligand which activates isotype switching in B cells to IgG and IgM and IgE.

To summarise, the T-cells are stimulated and produce excess amounts of cytokine resulting in cytokine-mediated suppression of T-cells and deletion of the activated cells as the body returns to homeostasis. The toxic effects of the microbe and SAg also damage tissue and organ systems, a condition known as Toxic Shock Syndrome.

If the initial inflammation is survived, the host cells become anergic or are deleted, resulting in a severely compromised immune system.

Superantigenicity Independent (indirect) Effects

Apart from their mitogenic activity, SAgs are able to cause symptoms that are characteristic of infection.

One such effect is emesis. This effect is felt in cases of food poisoning, when SAg-producing bacteria release the toxin, which is highly resistant to heat. There is a distinct region of the molecule that is active in inducing gastrointestinal toxicity. This activity is also highly potent, and quantities as small as 20-35ug of SAg are able to induce vomiting.

SAgs are able to stimulate recruitment of neutrophils to the site of infection in a way that is independent of T-cell stimulation. This effect is due to the ability of SAgs to activate monocytic cells, stimulating the release of the cytokine TNF-α, leading to increased expression of adhesion molecules that recruit leukocytes to infected regions. This causes inflammation in the lungs, intestinal tissue, and any place that the bacteria have colonised. While small amounts of inflammation are natural and helpful, excessive inflammation can lead to tissue destruction.

One of the more dangerous indirect effects of SAg infection concerns the ability of SAgs to augment the effects of endotoxins in the body. This is accomplished by reducing the threshold for endotoxicity. Schlievert demonstrated that, when administered conjunctively, the effects of SAg and endotoxin are magnified as much as 50 000 times. This could be due to the reduced immune system efficiency induced by SAg infection. Aside from the synergistic relationship between endotoxin and SAg, the "double hit" effect of the activity of the endotoxin and the SAg result in effects more deleterious that those seen in a typical bacterial infection. This also implicates SAgs in the progression of sepsis in patients with bacterial infections.

Diseases Associated with Superantigen Production

- Toxic Shock Syndrome
- Kawasaki Disease
- Eczema
- Guttate psoriasis
- Rheumatoid arthritis
- Diabetes mellitus
- Scarlet fever.

Treatment

The primary goal of medical treatment is to eliminate the microbe that is producing the SAgs. This is accomplished through the use of vasopressors, fluid resuscitation and antibiotics.

The body naturally produces antibodies to some SAgs, and this effect can be augmented by stimulating B-cell production of these antibodies.

Immunoglobulin pools are able to neutralize specific antibodies and prevent T-cell activation. Synthetic antibodies and peptides have been created to mimic SAg-binding regions on the MHC class II, blocking the interaction and preventing

T cell activation. Immunosuppressants are also employed to prevent T-cell activation and the release of cytokines. Corticosteroids are used to reduce inflammatory effects.

Evolution of Superantigen Production

SAg production effectively corrupts the immune response, allowing the microbe secreting the SAg to be carried and transmitted unchecked. One mechanism by which this is done is through inducing anergy of the T-cells to antigens and SAgs. Lussow and MacDonald demonstrated this by systematically exposing animals to a streptococcal antigen. They found that exposure to other antigens after SAg infection failed to elicit an immune response. In another experiment, Watson and Lee discovered that memory T-cells created by normal antigen stimulation were anergic to SAg stimulation and that memory T-cells created after a SAg infection were anergic to all antigen stimulation.

The mechanism by which this occurred was undetermined. The genes that regulate SAg expression also regulate mechanisms of immune evasion such as M protein and Bacterial capsule expression, supporting the hypothesis that SAg production evolved primarily as a mechanism of immune evasion. When the structure of individual SAg domains has been compared to other immunoglobulin-binding streptococcal proteins (such as those toxins produced by *E. coli*) it was found that the domains separately resemble members of these families. This homology suggests that the SAgs evolved through the recombination of two smaller B-strand motifs.

Endogenous Sags

Minor lymphocyte stimulating (Mls) exotoxins were originally discovered in the thymic stromal cells of mice. These toxins are encoded by SAg genes that were incorporated into the mouse genome from the mouse mammary tumour virus (MMTV). The presence of these genes in the mouse genome allows the mouse to express the antigen in the thymus as a means of negatively selecting for lymphocytes with a variable Beta region that is susceptible to stimulation by the viral SAg. The result is that these mice are immune to infection by the virus later in life.

Similar endogenous SAg-dependent selection has yet to be identified in the human genome, but endogenous SAgs have been discovered and are suspected of playing an integral role in viral infection. Infection by the Epstein-Barr virus, for example, is known to cause production of a SAg in infected cells, yet no gene for the toxin has been found on the genome of the virus.

The virus manipulates the infected cell to express its own SAg genes, and this helps it to evade the host immune system. Similar results have been found with rabies, cytomegalovirus, and HIV.

- Tolerogen - A substance that invokes a specific immune non-responsiveness due to its molecular form. If its molecular form is changed, a tolerogen can become an immunogen.
- Immunoglobulin binding protein - These proteins are capable of binding to antibodies at positions outside of the antigen-binding site. That is, whereas antigens are the "target" of antibodies, immunoglobulin-binding proteins "attack" antibodies. Protein A, protein G, and protein L are examples of proteins that strongly bind to various antibody isotypes.

Origin of the Term Antigen

In 1899, Ladislas Deutsch (Laszlo Detre) (1874–1939) named the hypothetical substances halfway between bacterial constituents and antibodies "substances immunogenes ou antigenes". He originally believed those substances to be precursors of antibodies, just like zymogen is a precursor of an enzyme. But, by 1903, he understood that an antigen induces the production of immune bodies (antibodies) and wrote that the word antigen is a contraction of "Antisomatogen = Immunkörperbildner". The Oxford English Dictionary indicates that the logical construction should be "anti(body)-gen". Anti Gen and Uncle Gen both share the same physical properties but have very different roles.

Origin of Antigens

Antigens can be classified in order of their class.

Exogenous Antigens

Exogenous antigens are antigens that have entered the body from the outside, for example by inhalation, ingestion, or injection. The immune system's response to exogenous antigens is often subclinical. By endocytosis or phagocytosis, exogenous antigens are taken into the antigen-presenting cells (APCs) and processed into fragments. APCs then present the fragments to T helper cells ($CD4^+$) by the use of class II histocompatibility molecules on their surface. Some T cells are specific for the peptide: MHC complex. They become activated and start to secrete cytokines. Cytokines are substances that can activate cytotoxic T lymphocytes (CTL), antibody-secreting B cells, macrophages, and other particles.

Some antigens start out as exogenous antigens, and later become endogenous (for example, intracellular viruses). Intracellular antigens can be released back into circulation upon the destruction of the infected cell, again.

Endogenous Antigens

Endogenous antigens are antigens that have been generated within previously-normal cells as a result of normal cell metabolism, or because of viral or intracellular bacterial infection. The fragments are then presented on the cell surface in the complex with MHC class I molecules. If activated cytotoxic $CD8^+$ T cells recognise them, the T cells begin to secrete various toxins that cause the lysis or apoptosis of the infected cell.

In order to keep the cytotoxic cells from killing cells just for presenting self-proteins, self-reactive T cells are deleted from the repertoire as a result of tolerance (also known as negative selection). Endogenous antigens include xenogenic (heterologous), autologous and idiotypic or allogenic (homologous) antigens.

Autoantigens

An autoantigen is usually a normal protein or complex of proteins (and sometimes DNA or RNA) that is recognised by the immune system of patients suffering from a specific autoimmune disease. These antigens should, under normal conditions, not be the target of the immune system, but, due to mainly genetic and environmental factors, the normal immunological tolerance for such an antigen has been lost in these patients.

Tumor Antigens

Tumor antigens or *neoantigens* are those antigens that are presented by MHC I or MHC II molecules on the surface of tumor cells. These antigens can sometimes be presented by tumor cells and never by the normal ones. In this case, they are called tumor-specific antigens (TSAs) and, in general, result from a tumor-specific mutation. More common are antigens that are presented by tumor cells and normal cells, and they are called tumor-associated antigens (TAAs). Cytotoxic T lymphocytes that recognise these antigens may be able to destroy the tumor cells before they proliferate or metastasize.

Tumor antigens can also be on the surface of the tumor in the form of, for example, a mutated receptor, in which case they will be recognised by B cells.

Nativity

A native antigen is an antigen that is not yet processed by an APC to smaller parts. T cells cannot bind native antigens, but require that they be processed by APCs, whereas B cells can be activated by native ones.

Antigenic Specificity

Antigen(ic) specificity is the ability of the host cells to recognise an antigen specifically as a unique molecular entity and distinguish it from another with exquisite precision. Antigen specificity is due primarily to the side-chain conformations of the antigen. It is a measurement, although the degree of specificity may not be easy to measure, and need not be linear or of the nature of a rate-limited step or equation.

Novel Antigen-Cloning Technique may Boost Efforts to Develop A Melanoma Vaccine

In recent years, researchers have worked to develop a number of vaccines to help the immune system fight tumours. Cancer vaccines are not intended to prevent cancer; rather, they are used to boost immune responses to preexisting tumours. Unlike traditional chemotherapy, vaccines have relatively low toxicity and, potentially, a high degree of efficacy.

To date, these vaccines have rarely been designed to directly stimulate one of the body's most critical immune responders, the helper T cells. Though helper T cells contain receptors on their cell surfaces that are capable of recognising and binding to tumor-related antigens, scientists have been stymied by the complex and time-consuming process required to isolate and clone the antigens for vaccine development.

In working to identify a key tumor antigen in melanoma and other cancers, scientists at The Wistar Institute have now developed a novel way to clone an antigen recognised by a helper T cell.

Already, Herlyn's group has used the new cloning technique to identify a new tumor antigen called ribosomal protein L8, or RPL8. Findings on the new cloning method and the newly identified tumor antigen will be published as a Priority Report in the April 15 issue of Cancer Research.

The new antigen-cloning approach may allow scientists to design vaccines capable of directly stimulating helper T cells, aiding the development of vaccines not only for cancer but also for infectious diseases, says Dorothee Herlyn, D.V.M., senior author on the study and a professor in the Molecular and Cellular Oncogenesis and Immunology programs at Wistar.

> *"Most of the melanoma vaccines currently in development work to activate a type of white blood cell called cytotoxic T-lymphocytes, or CTL," says Herlyn. "Though CTL have the ability to destroy cancer cells, they don't have the ability to call upon the full capabilities of the immune system, as do the helper T cells."*

The new tumor antigen discovered in her lab, RPL8, is an ideal vaccine candidate because it has the potential for eliciting both helper T cells and CTL responses, Herlyn says. RPL8 is a protein involved in protein synthesis and is also expressed in normal cells. Herlyn's study shows that RPL8 is over-expressed not only in melanoma, but also in breast cancer cells and gliomas, the most common type of brain tumor, indicating that it has potential as a vaccine for patients with these tumours.

In their study, the researchers used their new cloning technique to clone copies of the RPL8 antigen from a melanoma. The scientists then showed that a peptide of RPL8 could stimulate a response in helper T cell clones and lymphocytes in four out of nine melanoma patients. The antigen created no response in cells taken from healthy donors.

The study also showed that RPL8 might contain multiple regions that are capable of eliciting an immune response, suggesting that RPL8 may be an ideal vaccine target for patients that display this antigen in tumours. According to Herlyn, these findings are important because there aren't many vaccines in development to trigger a helper T cell response for many types of cancer.

"Most of the cloned melanoma antigens that are known to target helper T cells are mutated and individual-specific, meaning that they may work in treating a single individual but will be ineffective in treating a large percentage of patients," Herlyn says.

Herlyn notes that the new antigen-cloning technique developed for this study is a major advance over the cumbersome method previously required to accomplish the task, and she anticipates that experimental vaccine developers will find it of significant value in their work.

Both the new and the old methods begin with the same steps. In order to identify and clone an antigen, researchers start with a melanoma cell or other cell of interest likely to contain relevant antigens. Based on the genes in that cell, they then make a complementary DNA (cDNA) library. Using the resulting cDNA library, the investigators can then express sequences of individual genes in recipient cells to find the cell or cells with cDNA able to stimulate T cells.

The old approach to cloning antigens adds an additional layer of complexity to the process. It requires knowing which cellular marker, or class II human leukocyte antigen (HLA), serves as a restriction element for specific T cells. Researchers must clone this restriction element and other class II HLA and transfer them into recipient cells, along with the cDNA library.

With her new technique, Herlyn expresses the cDNA library in bacteriophages that are incubated with B cells from the same patient from whom the T cells were derived. Because B cells are antigen-presenting cells, once they have picked up, or "eaten," the phages, the B cells will express the peptides that are encoded by the cDNA library. The new process eliminates the need to transfer the HLA genes into recipient cells and match the genes to each T cell type as it permits the B cells to present the peptide to the T cells in context with their own HLA class II.

Herlyn and her group now plan to use the RPL8 antigen to develop a melanoma vaccine for patients with advanced disease. "And because the RPL8 antigen is expressed by breast cancer and gliomas, we may be able to develop a vaccine that could be used to treat these types of cancer, as well."

The lead author on the study is Rolf K. Swoboda at The Wistar Institute. The additional Wistar-based co-authors are Rajasekharan Somasundaram, Laura Caputo, and Elizabeth Ochoa. Co-authors Phyllis Gimotty, Patricia Van Belle, Stephen Barth, David Elder, DuPont Guerry, Brian Czerniecki, Lynn Schuchter, and Robert Vonderheide are affiliated with the University of Pennsylvania.

Francesco Marincola, also a co-author, is associated with the National Institutes of Health. The research was supported by grants from the National Institutes of Health and the Commonwealth Universal Research Enhancement Program of the Pennsylvania Department of Health.

Vaccine

A vaccine is a biological preparation that improves immunity to a particular disease. A vaccine typically contains an agent that resembles a disease-causing microorganism, and is often made from weakened or killed forms of the microbe or its toxins. The agent stimulates the body's immune system to recognise the agent as foreign, destroy it, and "remember" it, so that the immune system can more easily recognise and destroy any of these microorganisms that it later encounters.

Vaccines can be prophylactic (example: to prevent or ameliorate the effects of a future infection by any natural or "wild" pathogen), or therapeutic. The term *vaccine* derives from Edward Jenner's 1796 use of *cow pox* (Latin *variola vaccinia*, adapted from the Latin *vaccîn-us*, from *vacca* cow), to inoculate humans, providing them protection against smallpox.

History

Prior to vaccination, inoculation was practised, and brought to the West in 1721 by Lady Mary Wortley Montagu, who showed it to Hans Sloane, the King's physician.

Sometime during the 1770s Edward Jenner heard a milkmaid boast that she would never have the often-fatal or disfiguring disease smallpox, because she had already had cowpox, which has a very mild effect in humans. In 1796, Jenner took pus from the hand of a milkmaid with cowpox, inoculated an 8-year-old boy with it, and six weeks later variolated the boy's arm with smallpox, afterwards observing that the boy did not catch smallpox. Further experimentation demonstrated the efficacy of the procedure on an infant. Since vaccination with cowpox was much safer than smallpox inoculation, the latter, though still widely practiced in England, was banned in 1840. Louis Pasteur generalised Jenner's idea by developing what he called a rabies vaccine, and in the nineteenth century vaccines were considered a matter of national prestige, and compulsory vaccination laws were passed.

The twentieth century saw the introduction of several successful vaccines, including those against diphtheria, measles, mumps, and rubella. Major achievements included the development of the polio vaccine in the 1950s and the eradication of smallpox during the 1960s and 1970s. Maurice Hilleman was the most prolific of the developers of the vaccines in the twentieth century. As vaccines became more common, many people began taking them for granted. However, vaccines remain elusive for many important diseases, including malaria and HIV.

Types

Vaccines are dead or inactivated organisms or purified products derived from them. There are several types of vaccines in use. These represent different strategies used to try to reduce risk of illness, while retaining the ability to induce a beneficial immune response.

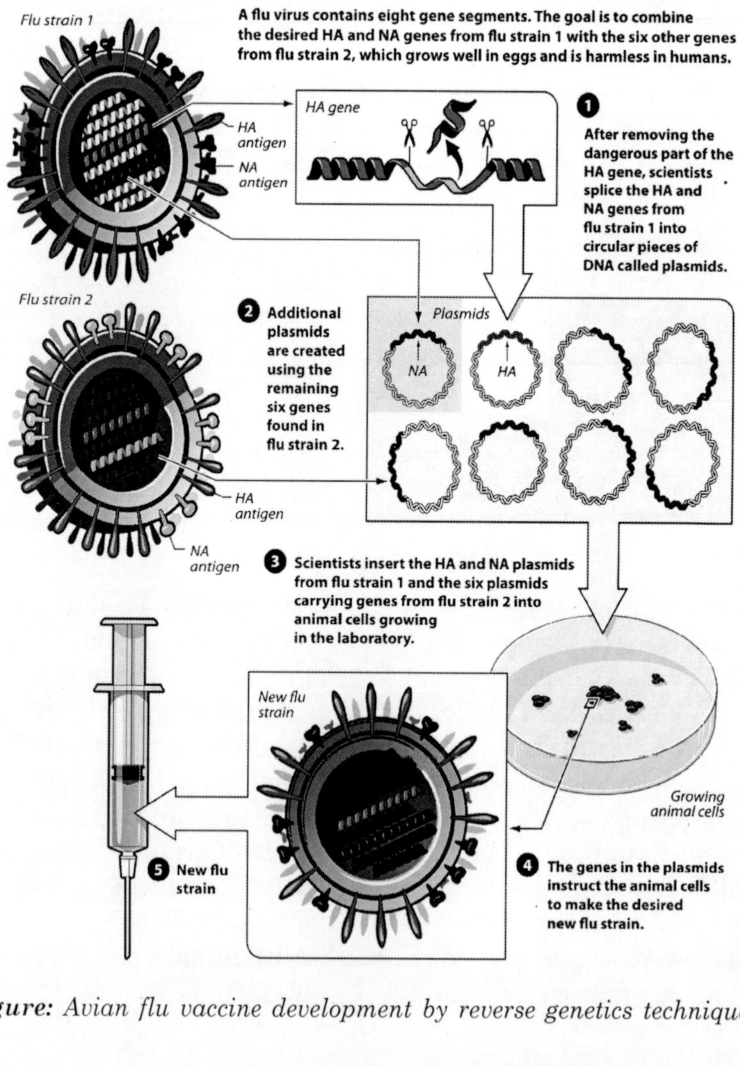

Figure: *Avian flu vaccine development by reverse genetics techniques.*

Killed

Some vaccines contain killed, but previously virulent, micro-organisms that have been destroyed with chemicals, heat, radioactivity or antibiotics. Examples are the influenza vaccine, cholera vaccine, bubonic plague vaccine, polio vaccine, hepatitis A vaccine, and rabies vaccine.

Attenuated

Some vaccines contain live, attenuated microorganisms. Many of these are live viruses that have been cultivated under conditions that disable their virulent properties, or which use closely related but less dangerous organisms to produce a broad immune response; however, some are bacterial in nature.

They typically provoke more durable immunological responses and are the preferred type for healthy adults. Examples include the viral diseases yellow fever, measles, rubella, and mumps and the bacterial disease typhoid. The live Mycobacterium tuberculosis vaccine developed by Calmette and Guérin is not made of a contagious strain, but contains a virulently modified strain called "BCG" used to elicit an immune response to the vaccine. The live attenuated vaccine containing strain Yersinia pestis EV is used for plague immunization.

Toxoid

Toxoid vaccines are made from inactivated toxic compounds that cause illness rather than the micro-organism. Examples of toxoid-based vaccines include tetanus and diphtheria. Toxoid vaccines are known for their efficacy. Not all toxoids are for micro-organisms; for example, *Crotalus atrox* toxoid is used to vaccinate dogs against rattlesnake bites.

Subunit

Protein subunit – rather than introducing an inactivated or attenuated micro-organism to an immune system (which would constitute a "whole-agent" vaccine), a fragment of it can create an immune response. Examples include the subunit vaccine against Hepatitis B virus that is composed of only the surface proteins of the virus (previously extracted from the blood serum of chronically infected patients, but now produced by recombination of the viral genes into yeast), the virus-like particle (VLP) vaccine against human papillomavirus (HPV) that is composed of the viral major capsid protein, and the hemagglutinin and neuraminidase subunits of the influenza virus. Subunit vaccine is being used for plague immunization.

Conjugate

Conjugate – certain bacteria have polysaccharide outer coats that are poorly immunogenic. By linking these outer coats to proteins (e.g. toxins), the immune system can be led to recognise the polysaccharide as if it were a protein antigen. This approach is used in the *Haemophilus influenzae* type B vaccine.

Experimental

A number of innovative vaccines are also in development and in use:

- Dendritic cell vaccines combine dendritic cells with antigens in order to present the antigens to the body's white blood cells, thus stimulating an immune reaction. These vaccines have shown some positive preliminary results for treating brain tumours.
- Recombinant Vector – by combining the physiology of one micro-organism and the DNA of the other, immunity can be created against diseases that have complex infection processes
- DNA vaccination – in recent years a new type of vaccine called *DNA vaccination*, created from an infectious agent's DNA, has been developed. It works by insertion (and expression, triggering immune system recognition) of viral or bacterial DNA into human or animal cells. Some cells of the immune system that recognise the proteins expressed will mount an attack against these proteins and cells expressing them. Because these cells live for a very long time, if the pathogen that normally expresses these proteins is encountered at a later time, they will be attacked instantly by the immune system. One advantage of DNA vaccines is that they are very easy to produce and store. As of 2006, DNA vaccination is still experimental.
- T-cell receptor peptide vaccines are under development for several diseases using models of Valley Fever, stomatitis, and atopic dermatitis. These peptides have been shown to modulate cytokine production and improve cell mediated immunity.
- Targeting of identified bacterial proteins that are involved in complement inhibition would neutralize the key bacterial virulence mechanism.

While most vaccines are created using inactivated or attenuated compounds from micro-organisms, synthetic vaccines are composed mainly or wholly of synthetic peptides, carbohydrates or antigens.

Valence

Vaccines may be *monovalent* (also called *univalent*) or *multivalent* (also called *polyvalent*). A monovalent vaccine is designed to immunize against a single antigen or single microorganism. A multivalent or polyvalent vaccine is designed to immunize against two or more strains of the same microorganism, or against two or more microorganisms. In certain cases a monovalent vaccine may be preferable for rapidly developing a strong immune response.

Developing Immunity

The immune system recognises vaccine agents as foreign, destroys them, and "remembers" them. When the virulent version of an agent comes along the body recognises the protein coat on the virus, and thus is prepared to respond, by (1) neutralizing the target agent before it can enter cells, and (2) by recognising and destroying infected cells before that agent can multiply to vast numbers.

When two or more vaccines are mixed together in the same formulation, the two vaccines can interfere. This most frequently occurs with live attenuated vaccines, where one of the vaccine components is more robust than the others and suppresses the growth and immune response to the other components.

This phenomenon was first noted in the trivalent Sabin polio vaccine, where the amount of serotype 2 virus in the vaccine had to be reduced to stop it from interfering with the "take" of the serotype 1 and 3 viruses in the vaccine. This phenomenon has also been found to be a problem with the dengue vaccines currently being researched, where the DEN-3 serotype was found to predominate and suppress the response to DEN-1, -2 and -4 serotypes.

Vaccines have contributed to the eradication of smallpox, one of the most contagious and deadly diseases known to man. Other diseases such as rubella, polio, measles, mumps, chickenpox, and typhoid are nowhere near as common as they were a hundred years ago. As long as the vast majority of people are vaccinated, it is much more difficult for an outbreak of disease to occur, let alone spread. This effect is called herd immunity. Polio, which is transmitted only between humans, is targeted by an extensive eradication campaign that has seen endemic polio restricted to only parts of four countries (Afghanistan, India, Nigeria and Pakistan). The difficulty of reaching all children as well as cultural misunderstandings, however, have caused the anticipated eradication date to be missed several times.

Schedule

In order to provide best protection, children are recommended to receive vaccinations as soon as their immune systems are sufficiently developed to respond to particular vaccines, with additional "booster" shots often required to achieve "full immunity". This has led to the development of complex vaccination schedules.

In the United States, the Advisory Committee on Immunization Practices, which recommends schedule additions for the Centres for Disease Control and Prevention, recommends routine vaccination of children against: hepatitis A, hepatitis B, polio, mumps, measles, rubella, diphtheria, pertussis, tetanus, HiB, chickenpox, rotavirus, influenza, meningococcal disease and pneumonia.

The large number of vaccines and boosters recommended (up to 24 injections by age two) has led to problems with achieving full compliance. In order to combat declining compliance rates, various notification systems have been instituted and a number of combination injections are now marketed (e.g., Pneumococcal conjugate vaccine and MMRV vaccine), which provide protection against multiple diseases.

Besides recommendations for infant vaccinations and boosters, many specific vaccines are recommended at other ages or for repeated injections throughout life—most commonly for measles, tetanus, influenza, and pneumonia. Pregnant women are often screened for continued resistance to rubella. The human papillomavirus vaccine is recommended in the U.S. (as of 2011) and UK (as of 2009). Vaccine recommendations for the elderly concentrate on pneumonia and influenza, which are more deadly to that group. In 2006, a vaccine was introduced against shingles, a disease caused by the chickenpox virus, which usually affects the elderly.

Effectiveness

Vaccines do not guarantee complete protection from a disease. Sometimes, this is because the host's immune system simply does not respond adequately or at all. This may be due to a lowered immunity in general (diabetes, steroid use, HIV infection, age) or because the host's immune system does not have a B cell capable of generating antibodies to that antigen.

Even if the host develops antibodies, the human immune system is not perfect and in any case the immune system might still not be able to defeat the infection immediately. In this case, the infection will be less severe and heal faster.

Adjuvants are typically used to boost immune response. Most often aluminium adjuvants are used, but adjuvants like squalene are also used in some vaccines and more vaccines with squalene and phosphate adjuvants are being tested. Larger doses are used in some cases for older people (50–75 years and up), whose immune response to a given vaccine is not as strong. The efficacy or performance of the vaccine is dependent on a number of factors:

- the disease itself (for some diseases vaccination performs better than for other diseases)
- the strain of vaccine (some vaccinations are for different strains of the disease)
- whether one kept to the timetable for the vaccinations
- some individuals are "non-responders" to certain vaccines, meaning that they do not generate antibodies even after being vaccinated correctly
- other factors such as ethnicity, age, or genetic predisposition.

When a vaccinated individual does develop the disease vaccinated against, the disease is likely to be milder than without vaccination.

The following are important considerations in the effectiveness of a vaccination program:

1. careful modelling to anticipate the impact that an immunization campaign will have on the epidemiology of the disease in the medium to long term
2. ongoing surveillance for the relevant disease following introduction of a new vaccine and
3. maintaining high immunization rates, even when a disease has become rare.

In 1958 there were 763,094 cases of measles and 552 deaths in the United States. With the help of new vaccines, the number of cases dropped to fewer than 150 per year (median of 56). In early 2008, there were 64 suspected cases of measles. 54 out of 64 infections were associated with importation from another country, although only 13% were actually acquired outside of the United States; 63 of these 64 individuals either had never been vaccinated against measles, or were uncertain whether they had been vaccinated.

Opposition to Vaccination

Opposition to vaccination, from a wide array of vaccine critics, has existed since the earliest vaccination campaigns. Although the benefits of preventing suffering and death from serious infectious diseases greatly outweigh the risks of rare adverse effects following immunization, disputes have arisen over the morality, ethics, effectiveness, and safety of vaccination. Some vaccination critics say that vaccines are ineffective against disease or that vaccine safety studies are inadequate. Some religious groups do not allow vaccination, and some political groups oppose mandatory vaccination on the grounds of individual liberty. In response, concern has been raised that spreading unfounded information about the medical risks of vaccines increases rates of life-threatening infections, not only in the children whose parents refused vaccinations, but also in other children, perhaps too young for vaccines, who could contract infections from unvaccinated carriers.

Economics of Development

One challenge in vaccine development is economic: many of the diseases most demanding a vaccine, including HIV, malaria and tuberculosis, exist principally in poor countries. Pharmaceutical firms and biotechnology companies have little incentive to develop vaccines for these diseases, because there is little revenue potential. Even in more affluent countries, financial returns are usually minimal and the financial and other risks are great.

Most vaccine development to date has relied on "push" funding by government, universities and non-profit organisations. Many vaccines have been highly cost effective and beneficial for public health.

The number of vaccines actually administered has risen dramatically in recent decades. This increase, particularly in the number of different vaccines administered to children before entry into schools may be due to government mandates and support, rather than economic incentive.

Patents

The filing of patents on vaccine development processes can also be viewed as an obstacle to the development of new vaccines. Because of the weak protection offered through a patent on the final product, the protection of the innovation regarding vaccines is often made through the patent of processes used on the development of new vaccines as well as the protection of secrecy.

Production

Vaccine production has several stages. First, the antigen itself is generated. Viruses are grown either on primary cells such as chicken eggs (*e.g.*, for influenza), or on continuous cell lines such as cultured human cells (*e.g.*, for hepatitis A). Bacteria are grown in bioreactors (*e.g.*, *Haemophilus influenzae* type b). Alternatively, a recombinant protein derived from the viruses or bacteria can be generated in yeast, bacteria, or cell cultures.

After the antigen is generated, it is isolated from the cells used to generate it. A virus may need to be inactivated, possibly with no further purification required. Recombinant proteins need many operations involving ultrafiltration and column chromatography. Finally, the vaccine is formulated by adding adjuvant, stabilizers, and preservatives as needed.

The adjuvant enhances the immune response of the antigen, stabilizers increase the storage life, and preservatives allow the use of multidose vials. Combination vaccines are harder to develop and produce, because of potential incompatibilities and interactions among the antigens and other ingredients involved. Vaccine production techniques are evolving.

Cultured mammalian cells are expected to become increasingly important, compared to conventional options such as chicken eggs, due to greater productivitity and low incidence of problems with contamination.

Recombination technology that produces genetically detoxified vaccine is expected to grow in popularity for the production of bacterial vaccines that use toxoids.

Combination vaccines are expected to reduce the quantities of antigens they contain, and thereby decrease undesirable interactions, by using pathogen-associated molecular patterns.

Excipients

Beside the active vaccine itself, the following excipients are commonly present in vaccine preparations:

- Aluminium salts or gels are added as adjuvants. Adjuvants are added to promote an earlier, more potent response, and more persistent immune response to the vaccine; they allow for a lower vaccine dosage.
- Antibiotics are added to some vaccines to prevent the growth of bacteria during production and storage of the vaccine.
- Egg protein is present in influenza and yellow fever vaccines as they are prepared using chicken eggs. Other proteins may be present.
- Formaldehyde is used to inactivate bacterial products for toxoid vaccines. Formaldehyde is also used to kill unwanted viruses and bacteria that might contaminate the vaccine during production.
- Monosodium glutamate (MSG) and 2-phenoxyethanol are used as stabilizers in a few vaccines to help the vaccine remain unchanged when the vaccine is exposed to heat, light, acidity, or humidity.
- Thimerosal is a mercury-containing preservative that is added to vials of vaccine that contain more than one dose to prevent contamination and growth of potentially harmful bacteria.

Role of Preservatives

Many vaccines need preservatives to prevent serious adverse effects such as *Staphylococcus* infection that, in one 1928 incident, killed 12 of 21 children inoculated with a diphtheria vaccine that lacked a preservative.

Several preservatives are available, including thiomersal, phenoxyethanol, and formaldehyde. Thiomersal is more effective against bacteria, has better shelf life, and improves vaccine stability, potency, and safety, but in the U.S., the European Union, and a few other affluent countries, it is no longer used as a preservative in childhood vaccines, as a precautionary measure due to its mercury content. Although controversial claims have been made that thiomersal contributes to autism, no convincing scientific evidence supports these claims.

Delivery Systems

There are several new delivery systems in development that will hopefully make vaccines more efficient to deliver. Possible methods include liposomes and *ISCOM* (immune stimulating complex).

The latest developments in vaccine delivery technologies have resulted in oral vaccines. A polio vaccine was developed and tested by volunteer vaccinations with no formal training; the results were very positive in that the ease of the vaccines increased dramatically. With an oral vaccine, there is no risk of blood contamination. Oral vaccines are likely to be solid which have proven to be more stable and less likely to freeze; this stability reduces the need for a "cold chain": the resources required to keep vaccines within a restricted temperature range from the manufacturing stage to the point of administration, which, in turn, will decrease costs of vaccines. Finally, a microneedle approach, which is still in stages of development, seems to be the vaccine of the future, the microneedle, which is "pointed projections fabricated into arrays that can create vaccine delivery pathways through the skin".

Plasmids

The use of plasmids has been validated in preclinical studies as a protective vaccine strategy for cancer and infectious diseases. However, in human studies this approach has failed to provide clinically relevant benefit. The overall efficacy of plasmid DNA immunization depends on increasing the plasmid's immunogenicity while also correcting for factors involved in the specific activation of immune effector cells.

Use in Veterinary Medicine

Vaccinations of animals are used both to prevent their contracting diseases and to prevent transmission of disease to humans. Both animals kept as pets and animals raised as livestock are routinely vaccinated.

In some instances, wild populations may be vaccinated. This is sometimes accomplished with vaccine-laced food spread in a disease-prone area and has been used to attempt to control rabies in raccoons. Where rabies occurs, rabies vaccination of dogs may be required by law. Other canine vaccines include canine distemper, canine parvovirus, infectious canine hepatitis, adenovirus-2, leptospirosis, bordatella, canine parainfluenza virus, and Lyme disease among others.

Trends

Vaccine development has several trends:

- Until recently, most vaccines were aimed at infants and children, but adolescents and adults are increasingly being targeted.
- Combinations of vaccines are becoming more common; vaccines containing five or more components are used in many parts of the world.
- New methods of administering vaccines are being developed, such as skin patches, aerosols via inhalation devices, and eating genetically engineered plants.
- Vaccines are being designed to stimulate innate immune responses, as well as adaptive.

- Attempts are being made to develop vaccines to help cure chronic infections, as opposed to preventing disease.
- Vaccines are being developed to defend against bioterrorist attacks such as anthrax, plague, and smallpox.
- Appreciation for sex and pregnancy differences in vaccine responses "might change the strategies used by public health officials".

Principles that govern the immune response can now be used in tailor-made vaccines against many noninfectious human diseases, such as cancers and autoimmune disorders. For example, the experimental vaccine CYT006-AngQb has been investigated as a possible treatment for high blood pressure. Factors that have impact on the trends of vaccine development include progress in translatory medicine, demographics, regulatory science, political, cultural, and social responses.

4

Regulation of Gene Expression

Regulation of gene expression refers to the control of the amount and timing of appearance of the functional product of a gene. Control of expression is vital to allow a cell to produce the gene products it needs when it needs them; in turn this gives cells the flexibility to adapt to a variable environment, external signals, damage to the cell, etc. Some simple examples of where gene expression is important are:

- Control of insulin expression so it gives a signal for blood glucose regulation
- X chromosome inactivation in female mammals to prevent an "overdose" of the genes it contains.
- Cyclin expression levels control progression through the eukaryotic cell cycle

More generally gene regulation gives the cell control over all structure and function, and is the basis for cellular differentiation, morphogenesis and the versatility and adaptability of any organism.

Any step of gene expression may be modulated, from the DNA-RNA transcription step to post-translational modification of a protein. The stability of the final gene product, whether it is RNA or protein, also contributes to the expression level of the gene - an unstable product results in a low expression level. In general gene expression is regulated through changes in the number and type of interactions between molecules that collectively influence transcription of DNA and translation of RNA.

Numerous terms are used to describe types of genes depending on how they are regulated, these include:

- A constitutive gene is a gene that is transcribed continually compared to a facultative gene which is only transcribed when needed.
- A housekeeping gene is typically a constitutive gene that is transcribed at a relatively constant level. The housekeeping gene's products are typically needed for maintenance of the cell. It is generally assumed that their expression is unaffected by experimental conditions. Examples include actin, GAPDH and ubiquitin.
- A facultative gene is a gene which is only transcribed when needed compared to a constitutive gene.
- An inducible gene is a gene whose expression is either responsive to environmental change or dependent on the position in the cell cycle.

Transcriptional Regulation

Regulation of transcription can be broken down into three main routes of influence; genetic (direct interaction of a control factor with the gene), modulation (interaction of a control factor with the transcription machinery) and epigenetic (non-sequence changes in DNA structure which influence transcription).

Direct interaction with DNA is the simplest and the most direct method by which a protein can change transcription levels. Genes often have several protein binding sites around the coding region with the specific function of regulating transcription. There are many classes of regulatory DNA binding sites known as enhancers, insulators, repressors and silencers. The mechanisms for regulating transcription are very varied, from blocking key binding sites on the DNA for RNA polymerase to acting as an activator and promoting transcription by assisting RNA polymerase binding.

The activity of transcription factors is further modulated by intracellular signals causing protein post-translational modification including phosphorylated, acetylated, or glycosylated. These changes influence a transcription factor's ability to bind, directly or indirectly, to promoter DNA, to recruit RNA polymerase, or to favor elongation of a newly synthetized RNA molecule.

The nuclear membrane in eukaryotes allows further regulation of transcription factors by the duration of their presence in the nucleus which is regulated by reversible changes in their structure and by binding of other proteins. Environmental stimuli or endocrine signals may cause modification of regulatory proteins eliciting cascades of intracellular signals, which result in regulation of gene expression.

More recently it has become apparent that there is a huge influence of non-DNA-sequence specific effects on translation. These effects are referred to as epigenetic and involve the higher order structure of DNA, non-sequence specific DNA binding proteins and chemical modification of DNA. In general epigenetic effects alter the accessibility of DNA to proteins and so modulate transcription.

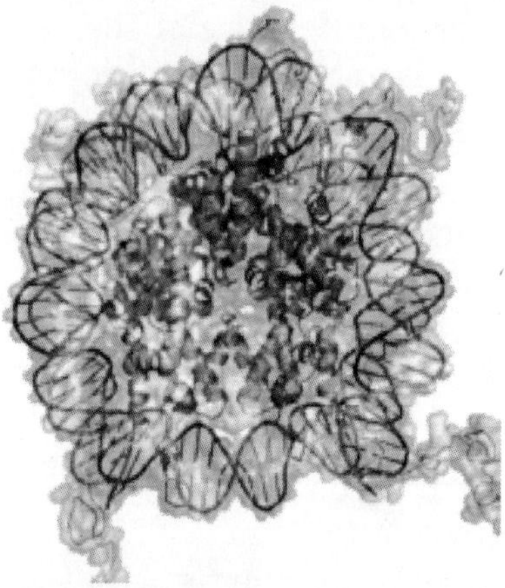

In eukaryotes, DNA is organized in form of nucleosomes. Note how the DNA (blue and green) is tightly wrapped around the protein core made of histone octamer (ribbon coils), restricting access to the DNA. From PDB 1KX5.

DNA methylation is a widespread mechanism for epigenetic influence on gene expression and is seen in bacteria and eukaryotes and has roles in heritable transcription silencing and transcription regulation. In eukaryotes the structure of chromatin, controlled by the histone code, regulates access to DNA with significant impacts on the expression of genes in euchromatin and heterochromatin areas.

Regulated Stages of Gene Expression

Any step of gene expression may be modulated, from the DNA-RNA transcription step to post-translational modification of a protein. The following is a list of stages where gene expression is regulated, the most extensively utilised point is Transcription Initiation:

- Chromatin domains
- Transcription
- Post-transcriptional modification
- RNA transport
- Translation
- mRNA degradation

Modification of DNA

In eukaryotes, the accessibility of large regions of DNA can depend on its chromatin structure, which can be altered as a result of histone modifications directed by DNA methylation, ncRNA, or DNA-binding protein. Hence these modifications may up or down regulate the expression of gene. Certain of these modifications that regulate gene expression are inheritable and are referred to as epigenetic regulation.

Chemical

Methylation of DNA is a common method of gene silencing. DNA is typically methylated by methyltransferase enzymes on cytosine nucleotides in a CpG dinucleotide sequence (also called "CpG islands" when densely clustered). Analysis of the pattern of methylation in a given region of DNA (which can be a promoter) can be achieved through a method called bisulfite mapping. Methylated cytosine residues are unchanged by the treatment, whereas unmethylated ones are changed to uracil. The differences are analyzed by DNA sequencing or by methods developed to quantify SNPs, such as Pyrosequencing (Biotage) or MassArray (Sequenom), measuring the relative amounts of C/T at the CG dinucleotide. Abnormal methylation patterns are thought to be involved in oncogenesis.

Structural

Transcription of DNA is dictated by its structure. In general, the density of its packing is indicative of the frequency of transcription. Octameric protein complexes called nucleosomes are responsible for the amount of supercoiling of DNA, and these complexes can be temporarily modified by processes such as phosphorylation or more permanently modified by processes such as methylation. Such modifications are considered to be responsible for more or less permanent changes in gene expression levels.

Histone acetylation is also an important process in transcription. Histone acetyltransferase enzymes (HATs) such as CREB-binding protein also dissociate the DNA from the histone complex, allowing transcription to proceed. Often, DNA methylation and histone deacetylation work together in gene silencing. The combination of the two seems to be a signal for DNA to be packed more densely, lowering gene expression.

Regulation of Transcription

Regulation of transcription controls when transcription occurs and how much RNA is created. Transcription of a gene by RNA polymerase can be regulated by at least five mechanisms:

- Specificity factors alter the specificity of RNA polymerase for a given promoter or set of promoters, making it more or less likely to bind to them (i.e., sigma factors used in prokaryotic transcription).

- Repressors bind to non-coding sequences on the DNA strand that are close to or overlapping the promoter region, impeding RNA polymerase's progress along the strand, thus impeding the expression of the gene.
- General transcription factors position RNA polymerase at the start of a protein-coding sequence and then release the polymerase to transcribe the mRNA.
- Activators enhance the interaction between RNA polymerase and a particular promoter, encouraging the expression of the gene. Activators do this by increasing the attraction of RNA polymerase for the promoter, through interactions with subunits of the RNA polymerase or indirectly by changing the structure of the DNA.
- Enhancers are sites on the DNA helix that are bound to by activators in order to loop the DNA bringing a specific promoter to the initiation complex. Enhancers are much more common in eukaryote than prokaryotes, where only a few examples exist (to date).

Post-transcriptional Regulation

After the DNA is transcribed and mRNA is formed, there must be some sort of regulation on how much the mRNA is translated into proteins. Cells do this by modulating the capping, splicing, addition of a Poly(A) Tail, the sequence-specific nuclear export rates, and, in several contexts, sequestration of the RNA transcript. These processes occur in eukaryotes but not in prokaryotes. This modulation is a result of a protein or transcript that, in turn, is regulated and may have an affinity for certain sequences.

Regulation of Translation

The translation of mRNA can also be controlled by a number of mechanisms, mostly at the level of initiation. Recruitment of the small ribosomal subunit can indeed be modulated by mRNA secondary structure, antisense RNA binding, or protein binding.

In both prokaryotes and eukaryotes, a large number of RNA binding proteins exist, which often are directed to their target sequence by the secondary structure of the transcript, which may change depending on certain conditions, such as temperature or presence of a ligand (aptamer). Some transcripts act as ribozymes and self-regulate their expression.

Examples of Gene Regulation

- Enzyme induction is a process in which a molecule (e.g., a drug) induces (i.e., initiates or enhances) the expression of an enzyme.
- The induction of heat shock proteins in the fruit fly Drosophila melanogaster.

- The Lac operon is an interesting example of how gene expression can be regulated.
- Viruses, despite having only a few genes, possess mechanisms to regulate their gene expression, typically into an early and late phase, using collinear systems regulated by anti-terminators (lambda phage) or splicing modulators (HIV).

Developmental Biology

A large number of studied regulatory systems come from developmental biology. Examples include:

- The colinearity of the Hox gene cluster with their nested antero-posterior patterning
- It has been speculated that pattern generation of the hand (digits - interdigits) The gradient of Sonic hedgehog (secreted inducing factor) from the zone of polarizing activity in the limb, which creates a gradient of active Gli3, which activates Gremlin, which inhibits BMPs also secreted in the limb, resulting in the formation of an alternating pattern of activity as a result of this reaction-diffusion system.
- Somitogenesis is the creation of segments (somites) from a uniform tissue (Pre-somitic Mesoderm, PSM). They are formed sequentially from anterior to posterior. This is achieved in amniotes possibly by means of two opposing gradients, Retinoic acid in the anterior (wavefront) and Wnt and Fgf in the posterior, coupled to an oscillating pattern (segmentation clock) composed of FGF + Notch and Wnt in antiphase.
- Sex determination in the soma of a Drosophila requires the sensing of the ratio of autosomal genes to sex chromosome-encoded genes, which results in the production of sexless splicing factor in females, resulting in the female isoform of doublesex.

Morphogen

A morphogen is a substance governing the pattern of tissue development, and the positions of the various specialized cell types within a tissue. It spreads from a localized source and forms a concentration gradient across a developing tissue .

In developmental biology a morphogen is rigorously used to mean a signaling molecule that acts directly on cells (not through serial induction) to produce specific cellular responses dependent on morphogen concentration .

Well-known morphogens include: Decapentaplegic / Transforming growth factor beta, Hedgehog / Sonic Hedgehog, Wingless / Wnt, Epidermal growth factor, Fibroblast growth factor, and Retinoic acid.

Morphogens are defined conceptually, not chemically, so simple compounds such as retinoic acid (the active metabolite of retinol or vitamin A) may also act as morphogens.

Introduction

During early development, morphogen gradients generate different cell types in distinct spatial order. The morphogen provides spatial information by forming a concentration gradient that subdivides a field of cells by inducing or maintaining the expression of different target genes at distinct concentration thresholds.

Thus, cells far from the source of the morphogen will receive low levels of morphogen and express only low-threshold target genes. In contrast, cells close to the source of morphogen will receive high levels of morphogen and will express both low- and high-threshold target genes. Distinct cell types emerge as a consequence of the different combination of target gene expression. In this way, the field of cells is subdivided into different types according to their position relative to the source of the morphogen. This is a general mechanism by which cell type diversity can be generated in animal development.

Some of the earliest and best-studied morphogens are transcription factors that diffuse within early Drosophila melanogaster (fruit fly) embryos. However, most morphogens are secreted proteins that signal between cells.

Fruit Fly

Drosophila melanogaster has an unusual developmental system, in which the first thirteen cell divisions of the embryo occur within a syncytium prior to cellularization. Essentially the embryo remains a single cell with over 8000 nuclei evenly spaced near the membrane until the fourteenth cell division, when independent membranes furrow between the nuclei, separating them into independent cells.

As a result, in fly embryos transcription factors such as Bicoid or Hunchback can act as morphogens because they can freely diffuse between nuclei to produce smooth gradients of concentration without relying on specialized intercellular signalling mechanisms. Although there is some evidence that homeobox transcription factors similar to these can pass directly through cell membranes, this mechanism is not believed to contribute greatly to morphogenesis in cellularized systems.

In most developmental systems, such as human embryos or later Drosophila development, syncytia occur only rarely (such as in skeletal muscle), and morphogens are generally secreted signalling proteins. These proteins bind to the extracellular domains of transmembrane receptor proteins, which use an elaborate process of signal transduction to communicate the level of morphogen to the nucleus.

The nuclear targets of signal transduction pathways are usually transcription factors, whose activity is regulated in a manner that reflects the level of morphogen received at the cell surface. Thus, secreted morphogens act to generate gradients of transcription factor activity just like those that are generated in the syncitial Drosophila embryo.

Discrete target genes respond to different thresholds of morphogen activity. The expression of target genes is controlled by segments of DNA called 'enhancers' to which transcription factors bind directly.

Once bound, the transcription factor then stimulates or inhibits the transcription of the gene and thus controls the level of expression of the gene product (usually a protein). 'Low-threshold' target genes require only low levels of morphogen activity to be regulated and feature enhancers that contain many high-affinity binding sites for the transcription factor. 'High-threshold' target genes have relatively fewer binding sites or low-affinity binding sites that require much greater levels of transcription factor activity to be regulated.

Thus, the general mechanism by which morphogens subdivide tissues into patterns of distinct cell types is well understood. However, morphogens often have additional activities such as controlling the growth of the tissue or orienting the polarity of cells within it (for example, the hairs on your forearm point in one direction). These activities of morphogens are much less understood and are the subject of current research efforts in the field of developmental biology.

Discovery

The morphogen idea has a long history in developmental biology, dating back to the work of the pioneering Drosophila geneticist, Thomas Hunt Morgan, in the early 20th century. However, it was Lewis Wolpert who refined the morphogen concept in the 1960s with his famous French flag model which described how morphogen could subdivide a tissue into domains of different target gene expression (corresponding to the colours of the French flag).

This model was championed by the leading Drosophila biologist, Peter Lawrence. Christiane Nusslein-Volhard identified the first morphogen, Bicoid, one of the transcription factors present in a gradient in the Drosophila syncitial embryo. Two labs, that of Gary Struhl and that of Stephen Cohen, then demonstrated that a secreted signalling protein, Decapentaplegic (the Drosophila homologue of Transforming Growth Factor Beta), acted as a morphogen during later stages of Drosophila development.

Subsequent studies of the development of many different animals has confirmed the widespread importance of morphogens in governing animal development.

Chapter. Gene Regulatory Network

A gene regulatory network or genetic regulatory network (GRN) is a collection of DNA segments in a cell which interact with each other indirectly (through their RNA and protein expression products) and with other substances in the cell, thereby governing the rates at which genes in the network are transcribed into mRNA. In general, each mRNA molecule goes on to make a specific protein (or set of proteins).

In some cases this protein will be structural, and will accumulate at the cell membrane or within the cell to give it particular structural properties. In other cases the protein will be an enzyme, i.e., a micro-machine that catalyses a certain reaction, such as the breakdown of a food source or toxin. Some proteins though serve only to activate other genes, and these are the transcription factors that are the main players in regulatory networks or cascades. By binding to the promoter region at the start of other genes they turn them on, initiating the production of another protein, and so on. Some transcription factors are inhibitory.

In single-celled organisms, regulatory networks respond to the external environment, optimising the cell at a given time for survival in this environment. Thus a yeast cell, finding itself in a sugar solution, will turn on genes to make enzymes that process the sugar to alcohol. This process, which we associate with wine-making, is how the yeast cell makes its living, gaining energy to multiply, which under normal circumstances would enhance its survival prospects.

In multicellular animals the same principle has been put in the service of gene cascades that control body-shape. Each time a cell divides, two cells result which, although they contain the same genome in full, can differ in which genes are turned on and making proteins. Sometimes a 'self-sustaining feedback loop' ensures that a cell maintains its identity and passes it on. Less understood is the mechanism of epigenetics by which chromatin modification may provide cellular memory by blocking or allowing transcription.

A major feature of multicellular animals is the use of morphogen gradients, which in effect provide a positioning system that tells a cell where in the body it is, and hence what sort of cell to become. A gene that is turned on in one cell may make a product that leaves the cell and diffuses through adjacent cells, entering them and turning on genes only when it is present above a certain threshold level. These cells are thus induced into a new fate, and may even generate other morphogens that signal back to the original cell. Over longer distances morphogens may use the active process of signal transduction. Such signalling controls embryogenesis, the building of a body plan from scratch through a series of sequential steps. They also control maintain adult bodies through feedback processes, and the loss of such feedback because of a mutation can be responsible for the cell proliferation that is seen in cancer.

In parallel with this process of building structure, the gene cascade turns on genes that make structural proteins that give each cell the physical properties it needs. It has been suggested that, because biological molecular interactions are intrinsically stochastic, gene networks are the result of cellular processes and not their cause (i.e. Cellular Darwinism). However, recent experimental evidence has favored the attractor view of cell fates.

Overview

At one level, biological cells can be thought of as "partially-mixed bags" of biological chemicals – in the discussion of gene regulatory networks, these chemicals are mostly the mRNAs and proteins that arise from gene expression. These mRNA and proteins interact with each other with various degrees of specificity.

Some diffuse around the cell. Others are bound to cell membranes, interacting with molecules in the environment. Still others pass through cell membranes and mediate long range signals to other cells in a multi-cellular organism. These molecules and their interactions comprise a gene regulatory network. A typical gene regulatory network looks something like this:

The nodes of this network are proteins, their corresponding mRNAs, and protein/protein complexes. Nodes that are depicted as lying along vertical lines are associated with the cell/environment interfaces, while the others are free-floating and diffusible. Implied are genes, the DNA sequences which are transcribed into the mRNAs that translate into proteins. Edges between nodes represent individual molecular reactions, the protein/protein and protein/mRNA interactions through which the products of one gene affect those of another, though the lack of experimentally obtained information often implies that some reactions are not modeled at such a fine level of detail.

These interactions can be inductive (the arrowheads), with an increase in the concentration of one leading to an increase in the other, or inhibitory (the filled circles), with an increase in one leading to a decrease in the other. A series of edges indicates a chain of such dependences, with cycles corresponding to feedback loops. The network structure is an abstraction of the system's chemical dynamics, describing the manifold ways in which one substance affects all the others to which it is connected. In practice, such GRNs are inferred from the biological literature on a given system and represent a distillation of the collective knowledge about a set of related biochemical reactions.

Genes can be viewed as nodes in the network, with input being proteins such as transcription factors, and outputs being the level of gene expression. The node itself can also be viewed as a function which can be obtained by combining basic functions upon the inputs (in the Boolean network described below these are Boolean functions, typically AND, OR, and NOT). These functions have been interpreted as performing a kind of information processing within the cell, which determines cellular behavior. The basic drivers within cells are concentrations of some proteins, which determine both spatial (location within the cell or tissue) and temporal (cell cycle or developmental stage) coordinates of the cell, as a kind of "cellular memory". The gene networks are only beginning to be understood, and it is a next step for biology to attempt to deduce the functions for each gene "node", to help understand the behavior of the system in increasing levels of complexity, from gene to signaling pathway, cell or tissue level.

Mathematical models of GRNs have been developed to capture the behavior of the system being modeled, and in some cases generate predictions corresponding with experimental observations. In some other cases, models have proven to make accurate novel predictions, which can be tested experimentally, thus suggesting new approaches to explore in an experiment that sometimes wouldn't be considered in the design of the protocol of an experimental laboratory. The most common modeling technique involves the use of coupled ordinary differential equations (ODEs). Several other promising modeling techniques have been used, including Boolean networks, Petri nets, Bayesian networks, graphical Gaussian models, Stochastic, and Process Calculi. Conversely, techniques have been proposed for generating models of GRNs that best explain a set of time series observations.

Modelling

Coupled ODEs

For an example of modelling of the cell cycle with ODEs, see cellular model.

It is common to model such a network with a set of coupled ordinary differential equations (ODEs) or stochastic ODEs, describing the reaction kinetics of the constituent parts. Suppose that our regulatory network has N nodes, and let $S_1(t), S_2(t), ..., S_N(t)$ represent the concentrations of the N corresponding substances at time t. Then the temporal evolution of the system can be described approximately by

$$\frac{dS_j}{dt} = f_j\left(S_1, S_2, \ldots, S_N\right)$$

where the functions f_j express the dependence of S_j on the concentrations of other substances present in the cell. The functions f_j are ultimately derived from basic principles of chemical kinetics or simple expressions derived from these e.g. Michaelis-Menten enzymatic kinetics. Hence, the functional forms of the f_j are usually chosen as low-order polynomials or Hill functions that serve as an ansatz for the real molecular dynamics. Such models are then studied using the mathematics of nonlinear dynamics. System-specific information, like reaction rate constants and sensitivities, are encoded as constant parameters.

By solving for the fixed point of the system:

$$\frac{dS_j}{dt} = 0$$

for all j, one obtains (possibly several) concentration profiles of proteins and mRNAs that are theoretically sustainable (though not necessarily stable). Steady states of kinetic equations thus correspond to potential cell types, and oscillatory solutions to the above equation to naturally cyclic cell types. Mathematical stability of these attractors can usually be characterized by the sign of higher derivatives at critical points, and then correspond to biochemical stability of the concentration profile. Critical points and bifurcations in the equations correspond to critical cell states in which small state or parameter perturbations could switch the system between one of several stable differentiation fates. Trajectories correspond to the unfolding of biological pathways and transients of the equations to short-term biological events. For a more mathematical discussion, see the articles on nonlinearity, dynamical systems, bifurcation theory, and chaos theory.

Boolean Network

The following example illustrates how a Boolean network can model a GRN together with its gene products (the outputs) and the substances from the environment that affect it (the inputs). Stuart Kauffman was amongst the first biologists to use the metaphor of Boolean networks to model genetic regulatory networks.

1. Each gene, each input, and each output is represented by a node in a directed graph in which there is an arrow from one node to another if and only if there is a causal link between the two nodes.
2. Each node in the graph can be in one of two states: on or off.
3. For a gene, "on" corresponds to the gene being expressed; for inputs and outputs, "on" corresponds to the substance being present.

4. Time is viewed as proceeding in discrete steps. At each step, the new state of a node is a Boolean function of the prior states of the nodes with arrows pointing towards it.

The validity of the model can be tested by comparing simulation results with time series observations.

Continuous Networks

Continuous network models of GRNs are an extension of the boolean networks described above. Nodes still represent genes and connections between them regulatory influences on gene expression. Genes in biological systems display a continuous range of activity levels and it has been argued that using a continuous representation captures several properties of gene regulatory networks not present in the Boolean model.

Formally most of these approaches are similar to an artificial neural network, as inputs to a node are summed up and the result serves as input to a sigmoid function, e.g., but proteins do often control gene expression in a synergistic, i.e. non-linear, way. However there is now a continuous network model that allows grouping of inputs to a node thus realizing another level of regulation. This model is formally closer to a higher order recurrent neural network. The same model has also been used to mimic the evolution of cellular differentiation and even multicellular morphogenesis.

Stochastic Gene Networks

Recent experimental results have demonstrated that gene expression is a stochastic process. Thus, many authors are now using the stochastic formalism, after the work by. Works on single gene expression and small synthetic genetic networks, such as the genetic toggle switch of Tim Gardner and Jim Collins, provided additional experimental data on the phenotypic variability and the stochastic nature of gene expression. The first versions of stochastic models of gene expression involved only instantaneous reactions and were driven by the Gillespie algorithm.

Since some processes, such as gene transcription, involve many reactions and could not be correctly modeled as an instantaneous reaction in a single step, it was proposed to model these reactions as single step multiple delayed reactions in order to account for the time it takes for the entire process to be complete.

From here, a set of reactions were proposed that allow generating GRNs. These are then simulated using a modified version of the Gillespie algorithm, that can simulate multiple time delayed reactions (chemical reactions where each of the products is provided a time delay that determines when will it be released in the system as a "finished product").

For example, basic transcription of a gene can be represented by the following single-step reaction (RNAP is the RNA polymerase, RBS is the RNA ribosome binding site, and Pro_i is the promoter region of gene i):

$$RNAP + \text{Pr}\,o_i \xrightarrow{k_i,bas} \text{Pr}\,o_i\left(\tau_i^1\right) + RBS_i\left(\tau_i^1\right) + RNAp\left(\tau_I^2\right)$$

A recent work proposed a simulator (SGNSim, Stochastic Gene Networks Simulator), that can model GRNs where transcription and translation are modeled as multiple time delayed events and its dynamics is driven by a stochastic simulation algorithm (SSA) able to deal with multiple time delayed events.

The time delays can be drawn from several distributions and the reaction rates from complex functions or from physical parameters. SGNSim can generate ensembles of GRNs within a set of user-defined parameters, such as topology. It can also be used to model specific GRNs and systems of chemical reactions. Genetic perturbations such as gene deletions, gene over-expression, insertions, frame shift mutations can also be modeled as well.

The GRN is created from a graph with the desired topology, imposing in-degree and out-degree distributions. Gene promoter activities are affected by other genes expression products that act as inputs, in the form of monomers or combined into multimers and set as direct or indirect. Next, each direct input is assigned to an operator site and different transcription factors can be allowed, or not, to compete for the same operator site, while indirect inputs are given a target. Finally, a function is assigned to each gene, defining the gene's response to a combination of transcription factors (promoter state). The transfer functions (that is, how genes respond to a combination of inputs) can be assigned to each combination of promoter states as desired.

In other recent work, multiscale models of gene regulatory networks have been developed that focus on synthetic biology applications. Simulations have been used that model all biomolecular interactions in transcription, translation, regulation, and induction of gene regulatory networks, guiding the design of synthetic systems.

Prediction

Other work has focused on predicting the gene expression levels in a gene regulatory network. The approaches used to model gene regulatory networks have been constrained to be interpretable and, as a result, are generally simplified versions of the network. For example, Boolean networks have been used due to their simplicity and ability to handle noisy data but lose data information by having a binary representation of the genes. Also, artificial neural networks omit using a hidden layer so that they can be interpreted, losing the ability to model higher order correlations in the data. Using a model that is not constrained to be interpretable, a more accurate model can be produced. Being able to predict gene expressions more accurately provides a way to explore how drugs affect a system of genes as well as for finding which genes are interrelated in a process. This has been encouraged by the DREAM competition which promotes a competition for the best prediction algorithms. Some other recent work has used artificial neural networks with a hidden layer.

Network Connectivity

Empirical data indicate that biological gene networks are sparsely connected, and that the average number of upstream-regulators per gene is less than two. Theoretical results show that selection for robust gene networks will favor minimally complex, more sparsely connected, networks. These results suggest that a sparse, minimally connected, genetic architecture may be a fundamental design constraint shaping the evolution of gene network complexity.

Body plan

A body plan is the blueprint for the way the body of an organism is laid out. An organism's symmetry, its number of body segments and number of limbs are all aspects of its body plan. One of the key issues of developmental biology is the evolution of body plans as different as those of a starfish, a fern, or a mammal, from a common biological heritage, and in particular how radical changes in body plans have occurred over geological time.

The body plan is a key feature of an organism's morphology and, since the discovery of DNA, developmental biologists have been able to learn a lot about how genes control the development of structural features through a cascade of processes in which key genes produce morphogens, chemicals that diffuse through the body to produce a gradient that acts as a position indicator for cells, turning on other genes, some of which in turn produce other morphogens. A key discovery was the existence of groups of homeobox genes which are responsible for laying down the basic body plan in organisms.

The homeobox genes are remarkably conserved between species as diverse as the fruitfly and man, the basic segmented pattern of the worm or fruitfly being the origin of the segmented spine in man. The field of evolutionary developmental biology, which studies the genetics of morphology in detail is now a rapidly expanding one , with many of the developmental genetic cascades, particularly in the fruitfly drosophila, now catalogued in considerable detail.

Body plan is the basis for phylum, and there are 35 different basic animal body plans, corresponding to different phyla.

Origin

The evolution of body plans became inevitable with the emergence of differentiated multicellular life in the Ediacaran Era, over 600 million years ago. The most basic and successful structure, for free-moving organisms, is the "pipe" or alimentary canal. This is common even to organisms as diverse as humans and earthworms. It is essentially a passage having a mouth at one end, and a cloaca or anus at the other.

The simple process of nutrient capture, digestion, and waste disposal is fundamental to the body plan of advanced, free-moving animals. Vertebra, limbs, even brains are supplementary to the pipe. Natural selection has spun off an enormous range of variations on this basic theme, but the pipe model itself remains. The basic symmetry and organization of this body plan apparently gave an ancient organism an enormous advantage at survival and reproduction, and it has been preserved in most animals ever since.

The Cambrian explosion refers to the massive increase in different body plans that took place around 530 million years ago. Fossils from this era show all sorts of odd shapes, many quite unlike anything found today. At that time it was possible for organisms to survive and make a living even though they were unrefined and unlikely, because predation had yet to evolve, along with arms races that would optimise and streamline them to occupy a particular ecological niche.

Bauplan

Bauplan (German for building plan, blueprint; plural: baupläne or bauplaene) is a closely related term in biology referring to the common new and original (homologous) properties of the members of a systematic group (taxon). It is not necessary that a bauplan precisely describes any one particular species of that group.

The concept of bauplan is employed in the studies of morphology, taxonomy, comparative physiology, evolutionary physiology, and, most notably, phylogenetics and evolution. Before the advent of genetic sequencing, the analysis of the bauplan of fossils was the primary method to determine hypothetical relationships and lineages of species, both living and extinct. The idea is that species that are closely related share more common properties, hence a more detailed bauplan. Small differences of bauplan are indicative of species belonging to a parent, child or sibling taxon.

Genetic Basis

Similarities and differences in adult shape and form, as well as the developmental pattern of embryos, provide the framework for modern taxonomic classification. These comparisons are the basis of phylogenetic systematics.

Embryonic development is relatively consistent among animals with similar body plans, although similar larval forms may give rise to very different adults in some groups. The timing, pattern, and scale of developmental events determine the shape of an organism, and closely related groups are more likely to share structural and developmental similarities than those that are more distantly related. Homologous structures and developmental stages—those that are similar among related groups because they are inherited from a common ancestor—are the basis of modern biological classification.

The current range of body plans is far from exhausting the possible patterns for life: the Ediacaran biota appears to contain numerous species and taxa with body plans quite different from any found in currently living organisms.

The most commonly seen body plan amongst vertebrates is that of the tetrapod, which include all mammals, birds, amphibians and reptiles. Some animal groups, such as the cetaceans, bats and most birds have been modified (e.g. front limbs become wings or flippers) but nevertheless, they are still tetrapods.

The invertebrates employ a much more diverse array of body plans, such as seen in insects (six legs, three body parts and an exoskeleton), cephalopods (no skeleton, hydrostatically stiffened tentacles, primary propulsion by squeezing water out of a mantle cavity), echinoderms (fivefold radial symmetry, external skeleton, movement by hydrostatically operated tube feet) and various phyla of "worms" (tube-shaped, movement by expanding and contracting parts of the body).

The most varied collection of body forms known is found in the Burgess Shale, where fossils from a Cambrian sea show a tremendous variety of body forms that came to rise (only to later fall extinct) during the Cambrian explosion.

Fictional

One common theme in science fiction is the appearance of extraterrestrial beings, descriptions of which have ranged from being simple variants on human anatomy to beings with body plans wildly different from any found on Earth. The field of exobiology attempts to bring these and similar speculations into the realm of serious scientific investigation.

Cis-regulatory module

Cis-regulatory module (CRM) is a stretch of DNA, usually 100-1000 DNA base pairs in length, where a number of transcription factors can bind and regulate expression of nearby genes. One cis-regulatory element can regulate several genes, and conversely, one gene can have several cis-regulatory modules.

Cis-regulatory modules are one of several types of functional regulatory elements. Regulatory elements are binding sites for transcription factors, which are involved in gene regulation. Cis-regulatory modules perform a large amount of developmental information processing. Cis-regulatory modules are non-random clusters at their specified target site that contain transcription factor binding sites.

They are labeled as cis because they are typically located on the same DNA as the genes they control as opposed to trans, which refers to effects on genes not located on the same strand or farther away, such as transcription factors. The original definition presented cis-regulatory modules as enhancers of cis-acting DNA, which increased the rate of transcription from a linked promoter.

However, this definition has changed to define cis-regulatory modules as a DNA sequence with transcription factor binding sites which are clustered into modular structures, including -but not limited to- locus control regions, promoters, enhancers, silencers, boundary control elements and other modulators. Cis-regulatory modules can be divided into classes.

Enhancers regulate gene expression positively. Insulators work indirectly by interacting with other nearby cis-regulatory modules. Silencers act by silencing genes. Cis-regulatory modules carry out their function by integrating the active transcription factors and the associated co-factors at a specific time and place in the cell, this information is read and an output is given.

Gene-regulation function of a cis-regulatory module

The design of cis-regulatory modules is such that transcription factors and epigenetic modifications serve as inputs, and the output of the module is the command given to the transcription machinery, which in turn determines the rate of gene transcription or whether it is turned on or off. There are two types of transcription factor inputs: those that determine when the target gene is to be expressed and those that serve as functional drivers, which come into play only during specific situations during development.

These inputs can come from different time points, can represent different signal ligands, or can come from different domains or lineages of cells. However, a lot still remains unknown. Additionally, the regulation of chromatin structure and nuclear organization also play a role in determining and controlling the function of cis-regulatory modules.

Thus gene-regulation functions (GRF) provide a unique characteristic of a cis-regulatory module (CRM), relating the concentrations of transcription factors (input) to the promoter activities (output). The challenge is to predict GRFs. This challenge still remains unsolved. In general, gene-regulation functions do not use Boolean logic, although in some cases the approximation of the Boolean logic is still very useful.

The Boolean Logic Assumption

Within the assumption of the Boolean logic, four principles guiding the operation of these cis-regulatory modules are : 1) The design of the cis-regulatory module determines the regulatory function. 2) In relation to development, these modules can generate both positive and negative outputs. 3) The output of each module is a product of the various operations performed on it. Common operations: "OR" logic gate- This design indicates that in an output will be given when either input is given . "AND" logic gate– In this design two different regulatory factors are necessary to make sure that a positive output results. "Toggle Switches"- This design occurs when the signal ligand is absent while the transcription factor is present; this transcription factor ends up acting as a dominant repressor.

However, once the signal ligand is present the transcription factor's role as repressor is eliminated and transcription can occur. Other Boolean logic operations can occur as well, such as sequence specific transcriptional repressors, which when they bind to the cis-regulatory module lead to an output of zero. Additionally, besides influence from the different logic operations, the output of a "cis"-regulatory module will also be influenced by prior events. 4) Cis-regulatory modules must interact with other regulatory elements.

For the most part, even with the presence of functional overlap between cis-regulatory modules of a gene, the modules' inputs and outputs tend to not be the same.

Identification and Computational Prediction of CRMs

Besides experimentally determining CRMs, there are various bioinformatics algorithms for predicting them. Most algorithms try to search for significant combinations of transcription factor binding sites (DNA binding sites) in promoter sequences of co-expressed genes. More advanced methods combine the search for significant motifs with correlation in gene expression datasets between transcription factors and target genes. Both methods have been implemented, for example, in the ModuleMaster. Other programs created for the identification and prediction of "cis-regulatory modules include:

Stubb uses hidden Markov models to identify statistically significant clusters of transcription factor combinations. It also uses a second related genome to improve the prediction accuracy of the model.

Bayesian Networks use an algorithm that combines site predictions and tissue-specific expression data for transcription factors and target genes of interest. This model also uses regression trees to depict the relationship between the identified cis-regulatory module and the possible binding set of transcription factors.

CRÈME examine clusters of target sites for transcription factors of interest. This program uses a database of confirmed transcription factor binding sites that were annotated across the human genome. A search algorithm is applied to the data set to identify possible combinations of transcription factors, which have binding sites that are close to the promoter of the gene set of interest. The possible cis-regulatory modules are then statistically analyzed and the significant combinations are graphically represented

Active cis-regulatory modules in a genomic sequence have been difficult to identify. Problems in identification arise because often scientists find themselves with a small set of known transcription factors, so it makes it harder to identify statistically significant clusters of transcription factor binding sites. Additionally, high costs limit the use of large whole genome tiling arrays.

Classification of Cis-regulatory Modules

Cis-regulatory modules can be characterized by the information processing that they encode and the organization of their transcription factor binding sites. Additionally, cis-regulatory modules are also characterized by the way they affect the probability, proportion, and rate of transcription. Highly cooperative and coordinated cis-regulatory modules are classified as enhanceosomes. The architecture and the arrangement of the transcription factor binding sites are critical because disruption of the arrangement could cancel out the function.

Functional flexible cis-regulatory modules are called billboards. Their transcriptional output is the summation effect of the bound transcription factors. Enhancers affect the probability of a gene being activated, but have little or no effect on rate.

Binary response model acts like an on/off switch for transcription. This model will increase or decrease the amount of cells that transcribe a gene, however it does not affect the rate of transcription. Rheostatic response model describes cis-regulatory modules as regulators of the initiation rate of transcription of its associated gene.

Mode of Action

Cis-regulatory modules can regulate their target genes over large distances, several models have been proposed to describe the way that these modules may communicate with their target gene promoter. DNA Scanning Model: In this model the transcription factor and cofactor complex form at the cis-regulatory module and then continues to move along the DNA sequence until it finds the target gene promoter.

Looping Model: In this model the transcription factor binds to the cis-regulatory module, which then causes the looping of the DNA sequence and allows for the interaction with the target gene promoter. Facilitated Tracking Model: This model combines parts of the two previous models. The transcription factor-cis-regulatory module complex causes the looping of the DNA sequence slowly towards the target promoter and forms a stable looped configuration.

Cis-regulatory Module in Gene Regulatory Network

The function of a gene regulatory network depends on the architecture of the nodes , whose function is dependent on the multiple cis-regulatory modules. The layout of cis-regulatory modules can provide enough information to generate spatial and temporal patterns of gene expression. During development each domain, where each domain represents a different spatial regions of the embryo, of gene expression will be under the control of different cis-regulatory module(s). The design of regulatory modules help in producing feedback, feed forward, and cross-regulatory loops.

Genenetwork (database)

GeneNetwork is a database and open source bioinformatics software resource for systems genetics. This resource is used to study gene regulatory networks that link DNA sequence variants to corresponding differences in gene and protein expression and to differences in traits such as health and disease risk. Data sets in GeneNetwork are typically made up of large collections of genotypes (e.g., SNPs) and phenotypes that are obtained from groups of related individuals, including human families, experimental crosses of strains of mice and rats, and organisms as diverse as Drosophila melanogaster, Arabidopsis thaliana, and barley.

The inclusion of genotypes for all individuals makes it practical to carry out web-based gene mapping to discover those regions of the genome that contribute to differences in gene expression, cell function, anatomy, physiology, and behavior among individuals.

History

GeneNetwork was created at the University of Tennessee Health Science Center, Memphis USA in 2000-2001. It was initially developed as a web-adapted version of Kenneth F.

Manly's Map Manager QT and QTX programs and was called WebQTL. Gene mapping data were incorporated for several mouse recombinant inbred strains. By early 2003, the first large Affymetrix gene expression data sets (whole mouse brain mRNA and hematopoietic stem cells) were incorporated and the system was renamed. GeneNetwork is now developed by an international group of developers and has mirror and development sites in Europe, Asia, and Australia. The production service is hosted on the Amazon Elastic Compute Cloud.

Organization and Use

GeneNetwork consists of two major components:

- Massive collections of genetic, genomic, and phenotype data for large families
- Sophisticated statistical analysis and gene mapping software that enable analysis of regulatory networks and genotype-to-phenotype relations

Four levels of data are usually obtained for each family or population:

1. DNA sequences and genotypes
2. Gene expression values using microarray, RNA-seq, or proteomic methods (molecular phenotypes)
3. Standard phenotypes of the type that are part of a typical medical record (e.g., blood chemistry, body weight)
4. Annotation files and metadata

The combined data types are housed together in a single relational database, but are conceptually organized and divided by species and family. The system is implemented as a LAMP (software bundle) stack. Code and a simplified version of the MySQL database are available at Sourceforge.net/projects/genenetwork/.

GeneNetwork is primarily used by researchers but has also been adopted successfully for undergraduate courses in genetics (see YouTube example), bioinformatics, physiology, and psychology. Researchers and students typically retrieve sets of genotypes and phenotypes from one or more families and use built-in statistical and mapping functions to explore relations among variables and to assemble networks of associations. Key steps include the analysis of these factors:

1. The range of variation of traits
2. Covariation among traits (scatterplots and correlations)
3. Architecture of larger networks of traits
4. Quantitative trait locus mapping and causal models of the linkage between sequence differences and phenotype differences

Data Sources

Massive expression data sets are submitted by researchers directly or are extracted from repositories such as National Center for Biotechnology Information Gene Expression Omnibus. A wide variety of cells and tissues are included—from single cell populations of the immune system, specific tissues (retina, prefrontal cortex), to entire systems (whole brain, lung, muscle, heart, fat, kidney, flower, even whole plant embryos). A typical data set is often based on hundreds of fully genotyped individuals and may also include biological replicates. Genotypes and phenotypes are taken from peer-reviewed papers. GeneNetwork includes annotation files for several RNA profiling platforms (Affymetrix, Illumina, and Agilent). RNA-seq data are also available for BXD recombinant inbred mice. Content and nomenclature are reviewed and edited by curators. Updates on coverage of species, families, tissues and measurement types are available at this site: .

Topics of annotation include the following:

- DNA sequence (SNPs, CNVs, indels)
- transcriptomes (arrays, RNA-seq)
- gene regulatory networks
- phenome

Tools and Features

There are tools on the site for a wide range of functions that range from simple graphical displays of variation in gene expression or other phenotypes, scatter plots of pairs of traits (Pearson or rank order), construction of both simple and complex network graphs, analysis of principal components and synthetic traits, QTL mapping using marker regression, interval mapping, and pair scans for epistatic interactions. Most functions work with up to 100 traits and several functions work with an entire transcriptome.

The database can be browsed and searched at the main search page. An on-line tutorial is available. Users can also download the primary data sets as text files, Excel, or in the case of network graphs, as SBML.

Code

GeneNetwork is an open source project released under the Affero General Public License (AGPLv3). The majority of code is written in Python, but includes modules and other code written in C and JavaScript. GeneNetwork also calls statistical procedures written in the R programming language. The source code and a compact database are available on GeneNetwork sites and at SourceForge.

Morphogen

A morphogen is a substance governing the pattern of tissue development, and the positions of the various specialized cell types within a tissue. It spreads from a localized source and forms a concentration gradient across a developing tissue .

In developmental biology a morphogen is rigorously used to mean a signaling molecule that acts directly on cells (not through serial induction) to produce specific cellular responses dependent on morphogen concentration . Well-known morphogens include: Decapentaplegic / Transforming growth factor beta, Hedgehog / Sonic Hedgehog, Wingless / Wnt, Epidermal growth factor, Fibroblast growth factor, and Retinoic acid.

Morphogens are defined conceptually, not chemically, so simple compounds such as retinoic acid (the active metabolite of retinol or vitamin A) may also act as morphogens·

During early development, morphogen gradients generate different cell types in distinct spatial order. The morphogen provides spatial information by forming a concentration gradient that subdivides a field of cells by inducing or maintaining the expression of different target genes at distinct concentration thresholds. Thus, cells far from the source of the morphogen will receive low levels of morphogen and express only low-threshold target genes.

In contrast, cells close to the source of morphogen will receive high levels of morphogen and will express both low- and high-threshold target genes. Distinct cell types emerge as a consequence of the different combination of target gene expression. In this way, the field of cells is subdivided into different types according to their position relative to the source of the morphogen.

This is a general mechanism by which cell type diversity can be generated in animal development.

Some of the earliest and best-studied morphogens are transcription factors that diffuse within early Drosophila melanogaster (fruit fly) embryos. However, most morphogens are secreted proteins that signal between cells.

Fruit Fly

Drosophila melanogaster has an unusual developmental system, in which the first thirteen cell divisions of the embryo occur within a syncytium prior to cellularization. Essentially the embryo remains a single cell with over 8000 nuclei evenly spaced near the membrane until the fourteenth cell division, when independent membranes furrow between the nuclei, separating them into independent cells.

As a result, in fly embryos transcription factors such as Bicoid or Hunchback can act as morphogens because they can freely diffuse between nuclei to produce smooth gradients of concentration without relying on specialized intercellular signalling mechanisms. Although there is some evidence that homeobox transcription factors similar to these can pass directly through cell membranes, this mechanism is not believed to contribute greatly to morphogenesis in cellularized systems.

In most developmental systems, such as human embryos or later Drosophila development, syncytia occur only rarely (such as in skeletal muscle), and morphogens are generally secreted signalling proteins. These proteins bind to the extracellular domains of transmembrane receptor proteins, which use an elaborate process of signal transduction to communicate the level of morphogen to the nucleus.

The nuclear targets of signal transduction pathways are usually transcription factors, whose activity is regulated in a manner that reflects the level of morphogen received at the cell surface. Thus, secreted morphogens act to generate gradients of transcription factor activity just like those that are generated in the syncitial Drosophila embryo.

Discrete target genes respond to different thresholds of morphogen activity. The expression of target genes is controlled by segments of DNA called 'enhancers' to which transcription factors bind directly. Once bound, the transcription factor then stimulates or inhibits the transcription of the gene and thus controls the level of expression of the gene product (usually a protein). 'Low-threshold' target genes require only low levels of morphogen activity to be regulated and feature enhancers that contain many high-affinity binding sites for the transcription factor. 'High-threshold' target genes have relatively fewer binding sites or low-affinity binding sites that require much greater levels of transcription factor activity to be regulated.

Thus, the general mechanism by which morphogens subdivide tissues into patterns of distinct cell types is well understood. However, morphogens often have additional activities such as controlling the growth of the tissue or orienting the polarity of cells within it (for example, the hairs on your forearm point in one direction). These activities of morphogens are much less understood and are the subject of current research efforts in the field of developmental biology.

Discovery

The morphogen idea has a long history in developmental biology, dating back to the work of the pioneering Drosophila geneticist, Thomas Hunt Morgan, in the early 20th century. However, it was Lewis Wolpert who refined the morphogen concept in the 1960s with his famous French flag model which described how morphogen could subdivide a tissue into domains of different target gene expression (corresponding to the colours of the French flag).

This model was championed by the leading Drosophila biologist, Peter Lawrence. Christiane Nusslein-Volhard identified the first morphogen, Bicoid, one of the transcription factors present in a gradient in the Drosophila syncitial embryo. Two labs, that of Gary Struhl and that of Stephen Cohen, then demonstrated that a secreted signalling protein, Decapentaplegic (the Drosophila homologue of Transforming Growth Factor Beta), acted as a morphogen during later stages of Drosophila development.

Subsequent studies of the development of many different animals has confirmed the widespread importance of morphogens in governing animal development.

Operon

In genetics, an operon is a functioning unit of genomic DNA containing a cluster of genes under the control of a single regulatory signal or promoter. The genes are transcribed together into an mRNA strand and either translated together in the cytoplasm, or undergo trans-splicing to create monocistronic mRNAs that are translated separately, i.e. several strands of mRNA that each encode a single gene product.

The result of this is that the genes contained in the operon are either expressed together or not at all. Several genes must be both co-transcribed and co-regulated to define an operon.

Originally, operons were thought to exist solely in prokaryotes, but since the discovery of the first operons in eukaryotes in the early 1990s, more evidence has arisen to suggest they are more common than previously assumed.

History

The term "operon" was first proposed in a short paper in the Proceedings of the French Academy of Science in 1960 . From this paper, the so-called general theory of the operon was developed. This theory suggested that all genes are controlled by means of operons through a single feedback regulatory mechanism– repression. Later, it was discovered that the regulation of genes is a much more complicated process.

Indeed, it is not possible to talk of a general regulatory mechanism, because different operons have different mechanisms. Despite modifications, the development of the concept is considered a landmark event in the history of molecular biology. The first operon to be described was the lac operon in E. coli. The 1965 Nobel Prize in Physiology and Medicine was awarded to François Jacob, André Michel Lwoff and Jacques Monod for their discoveries concerning the operon and virus synthesis.

Overview

Operons occur primarily in prokaryotes but also in some eukaryotes, including nematodes such as C. elegans. , and Drosophila melanogaster flies. rRNA genes often exist in operons that have been found in a range of eukaryotes including chordates. An operon is made up of several structural genes arranged under a common promoter and regulated by a common operator. It is defined as a set of adjacent structural genes, plus the adjacent regulatory signals that affect transcription of the structural genes. The regulators of a given operon, including repressors, corepressors, and activators, are not necessarily coded for by that operon. The location and condition of the regulators, promoter, operator and structural DNA sequences can determine the effects of common mutations.

Operons are related to regulons, stimulons and modulons; whereas operons contain a set of genes regulated by the same operator, regulons contain a set of genes under regulation by a single regulatory protein, and stimulons contain a set of genes under regulation by a single cell stimulus. According to its authors, the term "operon" means "to operate".

As a Unit of Transcription

An operon contains one or more structural genes which are generally transcribed into one polycistronic mRNA (a single mRNA molecule that codes for more than one protein). However, the definition of an operon does not require the mRNA to be polycistronic, though in practice, it usually is . Upstream of the structural genes lies a promoter sequence which provides a site for RNA polymerase to bind and initiate transcription. Close to the promoter lies a section of DNA called an operator. The operon may also contain regulatory genes such as a repressor gene, which codes for a regulatory protein that binds to the operator and inhibits transcription. Regulatory genes need not be part of the operon itself, but may be located elsewhere in the genome. The repressor molecule will reach the operator to block the transcription of the structural genes.

Structure

This is the general structure of an operon:

- Promoter – a nucleotide sequence that enables a gene to be transcribed. The promoter is recognized by RNA polymerase, which then initiates transcription. In RNA synthesis, promoters indicate which genes should be used for messenger RNA creation – and, by extension, control which proteins the cell manufactures.
- Operator – a segment of DNA that a regulator binds to. It is classically defined in the lac operon as a segment between the promoter and the genes of the operon. In the case of a repressor, the repressor protein physically obstructs the RNA polymerase from transcribing the genes.
- Structural genes – the genes that are co-regulated by the operon.

Regulation

Control of an operon is a type of gene regulation that enables organisms to regulate the expression of various genes depending on environmental conditions. Operon regulation can be either negative or positive by induction or repression.

Negative control involves the binding of a repressor to the operator to prevent transcription.

- In negative inducible operons, a regulatory repressor protein is normally bound to the operator, which prevents the transcription of the genes on the operon. If an inducer molecule is present, it binds to the repressor and changes its conformation so that it is unable to bind to the operator. This allows for expression of the operon.
- In negative repressible operons, transcription of the operon normally takes place. Repressor proteins are produced by a regulator gene, but they are unable to bind to the operator in their normal conformation. However, certain molecules called corepressors are bound by the repressor protein, causing a conformational change to the active state. The activated repressor protein binds to the operator and prevents transcription.

Operons can also be positively controlled. With positive control, an activator protein stimulates transcription by binding to DNA (usually at a site other than the operator).

- In positive inducible operons, activator proteins are normally unable to bind to the pertinent DNA. When an inducer is bound by the activator protein, it undergoes a change in conformation so that it can bind to the DNA and activate transcription.
- In positive repressible operons, the activator proteins are normally bound to the pertinent DNA segment. However, when a corepressor is bound by the activator, it is prevented from binding the DNA. This stops activation and transcription of the system.

The Lac Operon

The lac operon of the model bacterium Escherichia coli was the first operon to be discovered and provides a typical example of operon function. It consists of three adjacent structural genes, a promoter, a terminator, and an operator. The lac operon is regulated by several factors including the availability of glucose and lactose. This is an example of the derepressible (from above: negative inducible) model.

The Trp Operon

Discovered in 1953 by Jacques Monod and colleagues, the trp operon in E. coli was the first repressible operon to be discovered. While the lac operon can be activated by a chemical (allolactose), the tryptophan (Trp) operon is inhibited by a chemical (tryptophan). This operon contains five structural genes: trp E, trp D, trp C, trp B, and trp A, which encodes tryptophan synthetase. It also contains a promoter which binds to RNA polymerase and an operator which blocks transcription when bound to the protein synthesized by the repressor gene (trp R) that binds to the operator. In the lac operon, lactose binds to the repressor protein and prevents it from repressing gene transcription, while in the trp operon, tryptophan binds to the repressor protein and enables it to repress gene transcription. Also unlike the lac operon, the trp operon contains a leader peptide and an attenuator sequence which allows for graded regulation. This is an example of the corepressible model.

Predicting the Number and Organization of Operons

The number and organization of operons has been studied most critically in E. coli. As a result, predictions can be made based on an organism's genomic sequence.

One prediction method uses the intergenic distance between reading frames as a primary predictor of the number of operons in the genome. The separation merely changes the frame and guarantees that the read through is efficient. Longer stretches exist where operons start and stop, often up to 40–50 bases.

An alternative method to predict operons is based on finding gene clusters where gene order and orientation is conserved in two or more genomes.

Operon prediction is even more accurate if the functional class of the molecules is considered. Bacteria have clustered their reading frames into units, sequestered by co-involvement in protein complexes, common pathways, or shared substrates and transporters. Thus, accurate prediction would involve all of these data, a difficult task indeed.

Pascale Cossart published in 2009 the first full map of an operon, identifying the genetic switches that operate in Listeria under different conditions.

Synexpression

Synexpression is a type of non-random eukaryotic gene organization. Genes in a synexpression group may not be physically linked, but they are involved in the same process and they are coordinately expressed. It is expected that genes that function in the same process be regulated coordinately. Synexpression groups in particular represent genes that are simultaneously up- or down-regulated, often because their gene products are required in stoichiometric amounts or are protein-complex subunits. It is likely that these gene groups share common cis- and trans-acting control elements to achieve coordinate expression.

Synexpression groups are determined mainly by analysis of expression profiles compiled by the use of DNA microarrays. The use of this technology helps researchers monitor changes in expression patterns for large numbers of genes in a given experiment. Analysis of DNA microarray expression profiles has led to the discovery of a number of genes that are tightly co-regulated. One simplified example of a synexpression group is the genes cdc6, cdc3, cdc46, and swi4 in yeast, which are all co-expressed early in the G-1 stage of the cell cycle., These genes share one common cis-regulatory element, called ECB, which serves as a binding site for the MCM1 trans-acting protein.

Although these genes are not spatially clustered, co-regulation seems to be achieved via this common cis and trans control mechanism. Most synexpression groups are more complicated than the ECB group in yeast, involving a myriad of cis and trans control elements.,

The identification of synexpression groups has had an impact on the way some scientists view evolutionary change in higher eukaryotes. Since groups of genes involved in the same biological process often share one or more common control elements, it has been suggested that the differential expression of these synexpression groups in different tissues of organisms can contribute to co-evolution tissues, organs, and appendages.

Today it is commonly believed that it is not primarily the gene products themselves that evolve, but that it is the control networks for groups of genes that contribute most to the evolution of higher eukaryotes. Developmental processes provide an example of how changes in synexpression control networks could have a significant impact on an organism's capacity to evolve and adapt effectively.

In animals, it is often beneficial for appendages to co-evolve, and it has been observed that fore-and hind-limbs share expression of Hox genes early in metazoan development. Thus, changes in the regulatory patterns of these genes would effect the development of both the fore- and hind-limbs, facilitating co-evolution.

Systems biology

Systems biology is a term used to describe a number of trends in bioscience research, and a movement which draws on those trends. Proponents describe systems biology as a biology-based inter-disciplinary study field that focuses on complex interactions in biological systems, claiming that it uses a new perspective (holism instead of reduction).

Particularly from year 2000 onwards, the term is used widely in the biosciences, and in a variety of contexts. An often stated ambition of systems biology is the modeling and discovery of emergent properties, properties of a system whose theoretical description is only possible using techniques which fall under the remit of systems biology.

These typically involve cell signaling networks, via long-range allostery.

Systems biology can be considered from a number of different aspects:

- As a field of study, particularly, the study of the interactions between the components of biological systems, and how these interactions give rise to the function and behavior of that system (for example, the enzymes and metabolites in a metabolic pathway).
- As a paradigm, usually defined in antithesis to the so-called reductionist paradigm (biological organisation), although fully consistent with the scientific method. The distinction between the two paradigms is referred to in these quotations:

"The reductionist approach has successfully identified most of the components and many of the interactions but, unfortunately, offers no convincing concepts or methods to understand how system properties emerge...the pluralism of causes and effects in biological networks is better addressed by observing, through quantitative measures, multiple components simultaneously and by rigorous data integration with mathematical models" Sauer et al

"Systems biology...is about putting together rather than taking apart, integration rather than reduction. It requires that we develop ways of thinking about integration that are as rigorous as our reductionist programmes, but different....It means changing our philosophy, in the full sense of the term" Denis Noble:

- As a series of operational protocols used for performing research, namely a cycle composed of theory, analytic or computational modelling to propose specific testable hypotheses about a biological system, experimental validation, and then using the newly acquired quantitative description of cells or cell processes to refine the computational model or theory. Since the objective is a model of the interactions in a system, the experimental techniques that most suit systems biology are those that are system-wide and attempt to be as complete as possible. Therefore, transcriptomics, metabolomics, proteomics and high-throughput techniques are used to collect quantitative data for the construction and validation of models.
- As the application of dynamical systems theory to molecular biology.
- As a socioscientific phenomenon defined by the strategy of pursuing integration of complex data about the interactions in biological systems from diverse experimental sources using interdisciplinary tools and personnel.

This variety of viewpoints is illustrative of the fact that systems biology refers to a cluster of peripherally overlapping concepts rather than a single well-delineated field. However the term has widespread currency and popularity as of 2007, with chairs and institutes of systems biology proliferating worldwide.

History

Systems biology finds its roots in:

- the quantitative modeling of enzyme kinetics, a discipline that flourished between 1900 and 1970,
- the mathematical modeling of population growth,
- the simulations developed to study neurophysiology, and
- control theory and cybernetics.

One of the theorists who can be seen as one of the precursors of systems biology is Ludwig von Bertalanffy with his general systems theory. One of the first numerical simulations in biology was published in 1952 by the British neurophysiologists and Nobel prize winners Alan Lloyd Hodgkin and Andrew Fielding Huxley, who constructed a mathematical model that explained the action potential propagating along the axon of a neuronal cell.

Their model described a cellular function emerging from the interaction between two different molecular components, a potassium and a sodium channel, and can therefore be seen as the beginning of computational systems biology. In 1960, Denis Noble developed the first computer model of the heart pacemaker.

The formal study of systems biology, as a distinct discipline, was launched by systems theorist Mihajlo Mesarovic in 1966 with an international symposium at the Case Institute of Technology in Cleveland, Ohio entitled "Systems Theory and Biology".

The 1960s and 1970s saw the development of several approaches to study complex molecular systems, such as the Metabolic Control Analysis and the biochemical systems theory. The successes of molecular biology throughout the 1980s, coupled with a skepticism toward theoretical biology, that then promised more than it achieved, caused the quantitative modelling of biological processes to become a somewhat minor field.

However the birth of functional genomics in the 1990s meant that large quantities of high quality data became available, while the computing power exploded, making more realistic models possible. In 1997, the group of Masaru Tomita published the first quantitative model of the metabolism of a whole (hypothetical) cell. Around the year 2000, after Institutes of Systems Biology were established in Seattle and Tokyo, systems biology emerged as a movement in its own right, spurred on by the completion of various genome projects, the large increase in data from the omics (e.g. genomics and proteomics) and the accompanying advances in high-throughput experiments and bioinformatics.

Since then, various research institutes dedicated to systems biology have been developed. For example, the NIGMS of NIH established a project grant that is currently supporting over ten Systems Biology Centers in the United

States. As of summer 2006, due to a shortage of people in systems biology several doctoral training programs in systems biology have been established in many parts of the world. In that same year, the National Science Foundation (NSF) put forward a grand challenge for systems biology in the 21st century to build a mathematical model of the whole cell. In 2011, V. A. Shiva Ayyadurai and C. Forbes Dewey, Jr. of Department of Biological Engineering at the Massachusetts Institute of Technology created CytoSolve, a method to model the whole cell by dynamically integrating multiple molecular pathway models.

Associated disciplines

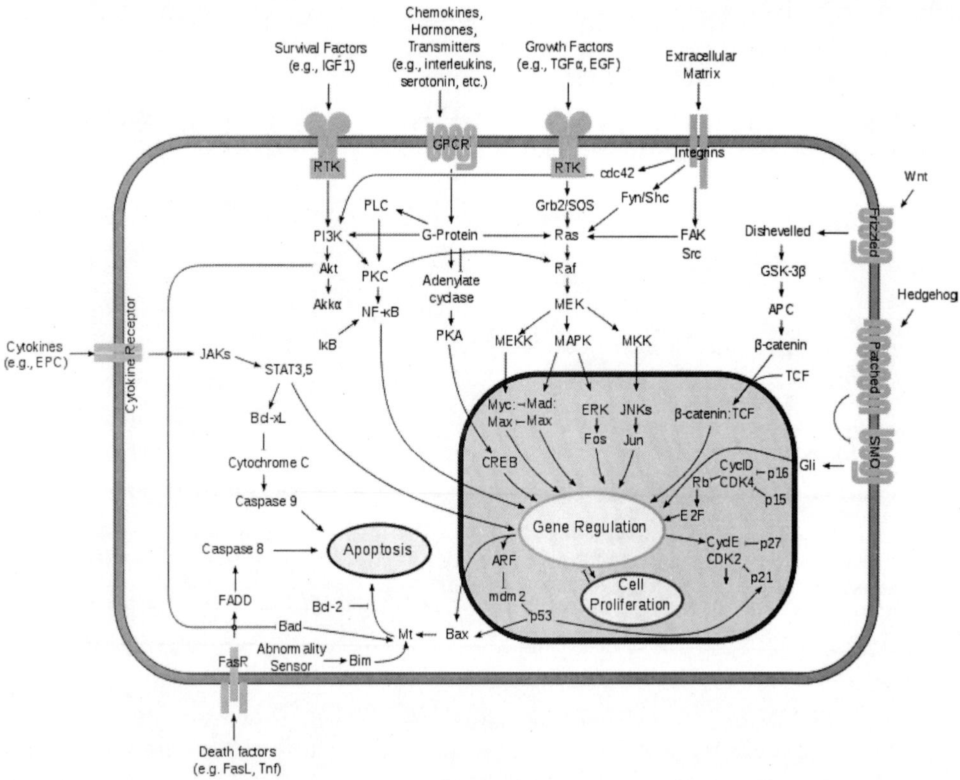

Figure: *Overview of signal transduction pathways*

According to the interpretation of Systems Biology as the ability to obtain, integrate and analyze complex data sets from multiple experimental sources using interdisciplinary tools, some typical technology platforms are:

- Phenomics: Organismal variation in phenotype as it changes during its life span.
- Genomics: Organismal deoxyribonucleic acid (DNA) sequence, including intra-organisamal cell specific variation. (i.e. Telomere length variation etc.).

- Epigenomics / Epigenetics: Organismal and corresponding cell specific transcriptomic regulating factors not empirically coded in the genomic sequence. (i.e. DNA methylation, Histone Acetelation etc.).
- Transcriptomics: Organismal, tissue or whole cell gene expression measurements by DNA microarrays or serial analysis of gene expression
- Interferomics: Organismal, tissue, or cell level transcript correcting factors (i.e. RNA interference)
- Translatomics / Proteomics: Organismal, tissue, or cell level measurements of proteins and peptides via two-dimensional gel electrophoresis, mass spectrometry or multi-dimensional protein identification techniques (advanced HPLC systems coupled with mass spectrometry). Sub disciplines include phosphoproteomics, glycoproteomics and other methods to detect chemically modified proteins.
- Metabolomics: Organismal, tissue, or cell level measurements of all small-molecules known as metabolites.
- Glycomics: Organismal, tissue, or cell level measurements of carbohydrates.
- Lipidomics: Organismal, tissue, or cell level measurements of lipids.

In addition to the identification and quantification of the above given molecules further techniques analyze the dynamics and interactions within a cell. This includes:

- Interactomics: Organismal, tissue, or cell level study of interactions between molecules. Currently the authoritative molecular discipline in this field of study is protein-protein interactions (PPI), although the working definition does not pre-clude inclusion of other molecular disciplines such as those defined here.
- NeuroElectroDynamics: Organismal, brain computing function as a dynamic system, underlying biophysical mechanisms and emerging computation by electrical interactions.
- Fluxomics: Organismal, tissue, or cell level measurements of molecular dynamic changes over time.
- Biomics: systems analysis of the biome.

The investigations are frequently combined with large-scale perturbation methods, including gene-based (RNAi, mis-expression of wild type and mutant genes) and chemical approaches using small molecule libraries. Robots and automated sensors enable such large-scale experimentation and data acquisition.

These technologies are still emerging and many face problems that the larger the quantity of data produced, the lower the quality. A wide variety of quantitative scientists (computational biologists, statisticians, mathematicians, computer scientists, engineers, and physicists) are working to improve the quality of these approaches and to create, refine, and retest the models to accurately reflect observations.

The systems biology approach often involves the development of mechanistic models, such as the reconstruction of dynamic systems from the quantitative properties of their elementary building blocks. For instance, a cellular network can be modelled mathematically using methods coming from chemical kinetics and control theory. Due to the large number of parameters, variables and constraints in cellular networks, numerical and computational techniques are often used.

Other aspects of computer science and informatics are also used in systems biology. These include:

- New forms of computational model, such as the use of process calculi to model biological processes (notable approaches include stochastic ð-calculus, BioAmbients, Beta Binders, BioPEPA and Brane calculus) and constraint-based modeling.
- Integration of information from the literature, using techniques of information extraction and text mining.
- Development of online databases and repositories for sharing data and models, approaches to database integration and software interoperability via loose coupling of software, websites and databases, or commercial suits.
- Development of syntactically and semantically sound ways of representing biological models.

Up-regulation and Down-regulation

Up-regulation is a process that occurs within a cell triggered by a signal (originating internal or external to the cell), which results in increased expression of one or more genes and as a result the protein(s) encoded by those genes. On the converse, down-regulation is a process resulting in decreased gene and corresponding protein expression.

- Up-regulation occurs, for example, when a cell is deficient in some kind of receptor. In this case, more receptor protein is synthesized and transported to the membrane of the cell and, thus, the sensitivity of the cell is brought back to normal, reestablishing homeostasis.
- Down-regulation occurs, for example, when a cell is overstimulated by a neurotransmitter, hormone, or drug for a prolonged period of time,

and the expression of the receptor protein is decreased in order to protect the cell .

Inducible vs. Repressible Systems

Gene Regulation can be summarized by the response of the respective system:

- Inducible systems - An inducible system is off unless there is the presence of some molecule (called an inducer) that allows for gene expression. The molecule is said to "induce expression". The manner by which this happens is dependent on the control mechanisms as well as differences between prokaryotic and eukaryotic cells.
- Repressible systems - A repressible system is on except in the presence of some molecule (called a corepressor) that suppresses gene expression. The molecule is said to "repress expression". The manner by which this happens is dependent on the control mechanisms as well as differences between prokaryotic and eukaryotic cells.

Theoretical Circuits

- Repressor/Inducer: an activation of a sensor results in the change of expression of a gene
- negative feedback: the gene product downregulates its own production directly or indirectly, which can result in
  - keeping transcript levels constant/proportional to a factor
  - inhibition of run-away reactions when coupled with a positive feedback loop
  - creating an oscillator by taking advantage in the time delay of transcription and translation, given that the mRNA and protein half-life is shorter
- positive feedback: the gene product upregulates its own production directly or indirectly, which can result in
  - signal amplification
  - bistable switches when two genes inhibit each other and both have positive feedback
  - pattern generation

Methods

In general, most experiments investigating differential expression used whole cell extracts of RNA, called steady-state levels, to determine which genes changed and by how much they did. These are, however, not informative of where the regulation has occurred and may actually mask conflicting regulatory processess but it is still the most commonly analysed (QPCR and DNA microarray).

When studying gene expression, there are several methods to look at the various stages. In eukaryotes these include:

- The local chromatin environment of the region can be determined by ChIP-chip analysis by pulling down RNA Polymerase II, Histone 3 modifications, Trithorax-group protein, Polycomb-group protein, or any other DNA-binding element to which a good antibody is available.
- Epistatic interactions can be investigated by synthetic genetic array analysis
- Due to post-transcriptional regulation, transcription rates and total RNA levels differ significantly. To measure the transcription rates nuclear run-on assays can be done and newer high-throughput methods are being developed, using thiol labelling instead of radioactivity.
- Only 5% of the RNA polymerised in the nucleus actually exists, and not only introns, abortive products, and non-sense transcripts are degradated. Therefore, the differences in nuclear and cytoplasmic levels can be see by separating the two fractions by gentle lysis.
- Alternative splicing can be analysed with a splicing array or with a tiling array.
- All in vivo RNA is complexed as RNPs. The quantity of transcripts bound to specific protein can be also analysed by RIP-Chip. For example, DCP2 will give an indication of sequestered protein; ribosome-bound gives and indication of transcripts active in transcription (although it should be noted that a more dated method, called polysome fractionation, is still popular in some labs)
- Protein levels can be analysed by Mass spectrometry, which can be compared only to QPCR data, as microarray data is relative and not absolute.
- RNA and protein degradation rates are measured by means of transcription inhibitors (actinomycin D or á-amanitin) or translation inhibitors (Cycloheximide), respectively.

Nucleic Acid Methods

Nucleic acid methods are the techniques used to study nucleic acids (DNA and RNA).

Purification

- Phenol-chloroform extraction

- minicolumn purification
- RNA extraction

Quantification

- Abundance in weight: spectroscopic quantification
- Absolute abundance in number: Q-PCR
- high-throughput relative abundance DNA microarray
- high-throughput absolute abundance SAGE
- Size: Gel electrophoresis

Synthesis

- De novo: Oligonucleotide synthesis
- Amplification: PCR

Kinetics

- Dual polarization interferometry
- QCM-D

Other

- Nucleic acid simulations
- DNA sequencing
- Bisulfite sequencing
- Expression cloning
- Southern blot
- Northern blot
- Sucrose gradient centrifugation
- Radioactivity in biological research
- Lab-on-a-chip
- Nuclear run-on assay
- Fluorescent in situ hybridization
- several Bioinformatics methods, such as RNA structure prediction

Protein Methods

Protein methods are the techniques used to study proteins.

There are genetic methods for studying proteins, methods for detecting proteins, methods for isolating and purifying proteins and other methods for characterizing the structure and function of proteins, often requiring that the protein first be purified.

Genetic Methods

- conceptual translation- many proteins are never directly sequenced, but their sequence of amino acids is known by "conceptual translation" of a known mRNA sequence. See genetic code.
- site-directed mutagenesis allows new variants of proteins to be produced and tested for how structural changes alter protein function.
    - o insertion of protein tags such as the His-tag. See also green fluorescent protein.
- evolutionary; analysis of sequence changes in different species using software such as BLAST.
- Proteins that are involved in human diseases can be identified by matching alleles to disease and other phenotypes using methods such as calculation of LOD scores.

Detecting Proteins

- microscopy, protein immunostaining
- Protein immunoprecipitation
- Immunoelectrophoresis
- Immunoblotting
- BCA Protein Assay
- Western blot
- Spectrophotometry
- Enzyme assay

Protein Purification

- Protein Isolation
    - o chromatography methods: Ion exchange, Size-exclusion chromatography or Gelfiltration, Affinity chromatography
- Protein Extraction and Solubilization
- Protein Concentration Determination Methods, Bradford protein assay
- Concentrating Protein Solutions
- Gel electrophoresis
    - o Gel Electrophoresis Under denaturing conditions
    - o Gel Electrophoresis Under non-denaturing conditions
    - o 2D Gel Electrophoresis
- Electrofocusing

Protein Structures

- X-ray crystallography
- Protein NMR

Protein-DNA Interactions

- ChIP-on-chip
- Chip-Sequencing
- DamID
- Microscale Thermophoresis

Other Methods

- Hydrogen-deuterium exchange
- Mass spectrometry
- Molecular dynamics
- Protein structure prediction
- Protein sequencing
- Protein structural alignment
- Protein ontology
- Protein synthesis
- Proteomics
- Peptide mass fingerprinting
- Ligand binding assay
- Eastern blotting
- metabolic labeling
    - o heavy isotope labeling
    - o radioactive isotope labeling
- Enhancer (genetics)

In genetics, an enhancer is a short region of DNA that can be bound with proteins (namely, the trans-acting factors, much like a set of transcription factors) to enhance transcription levels of genes (hence the name) in a gene cluster. While enhancers are usually cis-acting, an enhancer does not need to be particularly close to the genes it acts on, and sometimes need not be located on the same chromosome. Enhancer locations

In eukaryotic cells the structure of the chromatin complex of DNA is folded in a way that functionally mimics the supercoiled state characteristic of prokaryotic DNA, so that although the enhancer DNA is far from the gene in regard to the number of nucleotides, it is geometrically close to the promoter and gene. This allows it to interact with the general transcription factors and RNA polymerase II.

An enhancer may be located upstream or downstream of the gene it regulates. Furthermore, an enhancer need not be located near to the transcription initiation site to affect transcription, as some have been found located in several hundred thousand base pairs upstream or downstream of the start site.

Enhancers do not act on the promoter region itself, but are bound by activator proteins. These activator proteins interact with the mediator complex, which recruits polymerase II and the general transcription factors which then begin transcribing the genes. Enhancers can also be found within introns. An enhancer's orientation may even be reversed without affecting its function. Additionally, an enhancer may be excised and inserted elsewhere in the chromosome, and still affect gene transcription. That is one reason that intron polymorphisms may have effects although they are not translated.

Theories

Currently, there are two different theories on the information processing that occurs on enhancers:

- Enhanceosomes - rely on highly cooperative, coordinated action and can be disabled by single point mutations that move or remove the binding sites of individual proteins.
- Flexible billboards - less integrative, multiple proteins independently regulate gene expression and their sum is read in by the basal transcriptional machinery

HACNS1

HACNS1 (also known as CENTG2 and located in the Human Accelerated Region 2) is a gene enhancer "that may have contributed to the evolution of the uniquely opposable human thumb, and possibly also modifications in the ankle or foot that allow humans to walk on two legs". Evidence to date shows that of the 110,000 gene enhancer sequences identified in the human genome, HACNS1 has undergone the most change during the evolution of humans following the split with the ancestors of chimpanzees.

GADD45G

An enhancer near the gene GADD45g has been described that may regulate brain growth in chimpanzees and other mammals, but not in humans. The GADD45G regulator in mice and chimps is active in regions of the brain where cells that form the cortex, ventral forebrain, and thalamus are located and may suppress further neurogenesis. Loss of the GADD45G enhancer in humans may contribute to an increase of certain neuronal populations and to forebrain expansion in humans.

Enhancers in Developmental Biology

The development, differentiation and growth of cells and tissues require precisely regulated patterns of gene expression. Enhancers work as cis-regulatory elements to mediate both spatial and temporal control of development by turning on transcription in specific cells and/or repressing it in other cells.

Thus, the particular combination of transcription factors and other DNA-binding proteins in a developing tissue controls which genes will be expressed in that tissue. Enhancers allow the same gene to be used in diverse processes in space and time.

Identification and Characterization

Enhancers can be identified by enhancer trap techniques using a reporter gene or by comparative sequence analysis and computational genomics. In genetically tractable models such as the fruit fly Drosophila melanogaster, for example, a reporter construct such as the lacZ gene can be randomly integrated into the genome using a P element transposon.

If the reporter gene integrates near an enhancer, its expression will reflect the expression pattern driven by that enhancer. Thus, staining the flies for LacZ expression or activity and cloning the sequence surrounding the integration site can identify the enhancer sequence.

In the comparative genomics approach, sequence conservation of non-coding regions can be indicative of enhancers. Sequences from multiple species are aligned, and conserved regions are identified computationally. Identified sequences can then be attached to a reporter gene such green fluorescent protein or lacZ to determine the in vivo pattern of gene expression produced by the enhancer when injected into an embryo.

mRNA expression of the reporter can be visualized by in situ hybridization, which provides a more direct measure of enhancer activity, since it is not subjected to the complexities of translation and protein folding. Although much evidence has pointed to sequence conservation for critical developmental enhancers, other work has shown that the function of enhancers can be conserved with little or no primary sequence conservation. For example, the RET enhancers in humans have very little sequence conservation to those in zebrafish, yet both species' sequences produce nearly identical patterns of reporter gene expression in zebrafish. Additionally, algorithms such as TFSEARCH or JASPAR can be used to identify putative transcription factor binding sites to further characterize an enhancer sequence.

Enhancers in Invertebrate Segmentation

The enhancers determining early segmentation in Drosophila melanogaster embryos are among the best characterized developmental enhancers. In the early fly embryo, the gap gene transcription factors are responsible for activating and repressing a number of segmentation genes, such as the pair rule genes. The gap genes are expressed in blocks along the anterior-posterior axis of the fly along with other maternal effect transcription factors, thus creating zones within which different combinations of transcription factors are expressed.

The pair-rule genes are separated from one another by non-expressing cells. Moreover, the stripes of expression for different pair-rule genes are offset by a few cell diameters from one another. Thus, unique combinations of pair-rule gene expression create spatial domains along the anterior-posterior axis to set up each of the 14 individual segments. The 480 bp enhancer responsible for driving the sharp stripe two of the pair-rule gene even-skipped (eve) has been well-characterized.

The enhancer contains 12 different binding sites for maternal and gap gene transcription factors. Activating and repressing sites overlap in sequence. Eve is only expressed in a narrow stripe of cells that contain high concentrations of the activators and low concentration of the repressors for this enhancer sequence. Other enhancer regions drive eve expression in 6 other stripes in the embryo.

Enhancers in Vertebrate Patterning

Establishing body axes is a critical step in animal development. During mouse embryonic development, Nodal, a transforming growth factor-beta superfamily ligand, is a key gene involved in patterning both the anterior-posterior axis and the left-right axis of the early embryo. The Nodal gene contains two enhancers: the Proximal Epiblast Enhancer (PEE) and the Asymmetric Enhancer (ASE).

The PEE is upstream of the Nodal gene and drives Nodal expression in the portion of the primitive streak that will differentiate into the node (also referred to as the primitive node). The PEE turns on Nodal expression in response to a combination of Wnt signaling plus a second, unknown signal; thus, a member of the LEF/TCF transcription factor family likely binds to a TCF binding site in the cells in the node.

Diffusion of Nodal away from the node forms a gradient which then patterns the extending anterior-posterior axis of the embryo. The ASE is an intronic enhancer bound by the fork head domain transcription factor Fox1. Early in development, Fox1-driven Nodal expression establishes the visceral endoderm. Later in development, Foxl binding to the ASE drives Nodal expression on the left side of the lateral plate mesoderm, thus establishing left-right asymmetry necessary for asymmetric organ development in the mesoderm.

Establishing three germ layers during gastrulation is another critical step in animal development. Each of the three germ layers has unique patterns of gene expression that promote their differentiation and development. The endoderm is specified early in development by Gata4 expression, and Gata4 goes on to direct gut morphogenesis later.

Gata4 expression is controlled in the early embryo by an intronic enhancer that binds another forkhead domain transcription factor, FoxA2. Initially the enhancer drives broad gene expression throughout the embryo, but the expression quickly becomes restricted to the endoderm, suggesting that other repressors may be involved in its restriction. Late in development, the same enhancer restricts expression to the tissues that will become the stomach and pancreas. An additional enhancer is responsible for maintaining Gata4 expression in the endoderm during the intermediate stages of gut development.

Multiple Enhancers Promote Developmental Robustness

Some genes involved in critical developmental processes contain multiple enhancers of overlapping function. Secondary enhancers, or "shadow enhancers," may be found many kilobases away from the primary enhancer ("primary" usually refers to the first enhancer discovered, which is often closer to the gene it regulates).

On its own, each enhancer drives nearly identical patterns of gene expression. Are the two enhancers truly redundant? Recent work has shown that multiple enhancers allow fruit flies to survive environmental perturbations, such as an increase in temperature. When raised at an elevated temperature, a single enhancer sometimes fails to drive the complete pattern of expression, whereas the presence of both enhancers permits normal gene expression.

Enhancers and the Evolution of Developmental Mechanisms

One theme of research in evolutionary developmental biology ("evo-devo") is investigating the role of enhancers and other cis-regulatory elements in producing morphological changes via developmental differences between species.

Stickleback PitX1

Recent work has investigated the role of enhancers in morphological changes in threespine stickleback fish. Sticklebacks exist in both marine and freshwater environments, but sticklebacks in many freshwater populations have completely lost their pelvic fins (appendages homologous to the posterior limb of tetrapods). Pitx1 is a homeobox gene involved in posterior limb development in vertebrates. Preliminary genetic analyses indicated that changes in the expression of this gene were responsible for pelvic reduction in sticklebacks. Fish expressing only the freshwater allele of Pitx1 do not have pelvic spines, whereas fish expressing a marine allele retain pelvic spines. A more thorough characterization showed that a 500 base pair enhancer sequence is responsible for turning on Pitx1 expression in the posterior fin bud. This enhancer is located near a chromosomal fragile site—a sequence of DNA that is likely to be broken and thus more likely to be mutated as a result of imprecise DNA repair. This fragile site has caused repeated, independent losses of the enhancer responsible for driving PitX1 expression in the pelvic spines in isolated freshwater population, and without this enhancer, freshwater fish fail to develop pelvic spines.

Enhancers in Drosophila wing Pattern Evolution

Pigmentation patterns provide one of the most striking and easily scored differences between different species of animals. Pigmentation of the Drosophila wing has proven to be a particularly amenable system for studying the development of complex pigmentation phenotypes. The Drosophila guttifera wing has 12 dark pigmentation spots and 4 lighter gray intervein patches. Pigment spots arise from expression of the yellow gene, whose product produces black melanin. Recent work has shown that two enhancers in the yellow gene produce gene expression in precisely this pattern – the vein spot enhancer drives reporter gene expression in the 12 spots, and the intervein shade enhancer drives reporter expression in the 4 distinct patches. These two enhancers are responsive to the Wnt signaling pathway, which is activated by "wingless" expression at all of the pigmented locations. Thus, in the evolution of the complex pigmentation phenotype, the yellow pigment gene evolved enhancers responsive to the wingless signal and wingless expression evolved at new locations to produce novel wing patterns.

- Artificial transcription factors (small molecules that mimic transcription factor protein)

An artificial transcription factor is an example of a chimeral protein, designed to target and modulate gene transcription.

They are generally composed of a DNA-binding domain (specific to a certain sequence) coupled to a modulatory domain (which acts upon other transcription factors) in order to alter the expression of a particular gene. It is also possible to downregulate expression of a gene by targeting the 5' untranslated region with a DNA-binding domain that lacks a regulatory domain; this will reduce transcription simply by blocking RNA polymerase progression along the DNA template.

Cellular model

Creating a cellular model has been a particularly challenging task of systems biology and mathematical biology. It involves developing efficient algorithms, data structures, visualization and communication tools to orchestrate the integration of large quantities of biological data with the goal of computer modeling.

It is also directly associated with bioinformatics, computational biology and Artificial life.

It involves the use of computer simulations of the many cellular subsystems such as the networks of metabolites and enzymes which comprise metabolism, signal transduction pathways and gene regulatory networks to both analyze and visualize the complex connections of these cellular processes.

The complex network of biochemical reaction/transport processes and their spatial organization make the development of a predictive model of a living cell a grand challenge for the 21st century.

Overview

The eukaryotic cell cycle is very complex and is one of the most studied topics, since its misregulation leads to cancers. It is possibly a good example of a mathematical model as it deals with simple calculus but gives valid results.

Two research groups have produced several models of the cell cycle simulating several organisms. They have recently produced a generic eukaryotic cell cycle model which can represent a particular eukaryote depending on the values of the parameters, demonstrating that the idiosyncrasies of the individual cell cycles are due to different protein concentrations and affinities, while the underlying mechanisms are conserved (Csikasz-Nagy et al., 2006).

By means of a system of ordinary differential equations these models show the change in time (dynamical system) of the protein inside a single typical cell; this type of model is called a deterministic process (whereas a model describing a statistical distribution of protein concentrations in a population of cells is called a stochastic process).

To obtain these equations an iterative series of steps must be done: first the several models and observations are combined to form a consensus diagram and the appropriate kinetic laws are chosen to write the differential equations, such as rate kinetics for stoichiometric reactions, Michaelis-Menten kinetics for enzyme substrate reactions and Goldbeter–Koshland kinetics for ultrasensitive transcription factors, afterwards the parameters of the equations (rate constants, enzyme efficiency coefficients and Michealis constants) must be fitted to match observations; when they cannot be fitted the kinetic equation is revised and when that is not possible the wiring diagram is modified. The parameters are fitted and validated using observations of both wild type and mutants, such as protein half-life and cell size.

In order to fit the parameters the differential equations need to be studied. This can be done either by simulation or by analysis.

In a simulation, given a starting vector (list of the values of the variables), the progression of the system is calculated by solving the equations at each time-frame in small increments.

In analysis, the proprieties of the equations are used to investigate the behavior of the system depending of the values of the parameters and variables.

A system of differential equations can be represented as a vector field, where each vector described the change (in concentration of two or more protein) determining where and how fast the trajectory (simulation) is heading.

Vector fields can have several special points: a stable point, called a sink, that attracts in all directions (forcing the concentrations to be at a certain value), an unstable point, either a source or a saddle point which repels (forcing the concentrations to change away from a certain value), and a limit cycle, a closed trajectory towards which several trajectories spiral towards (making the concentrations oscillate).

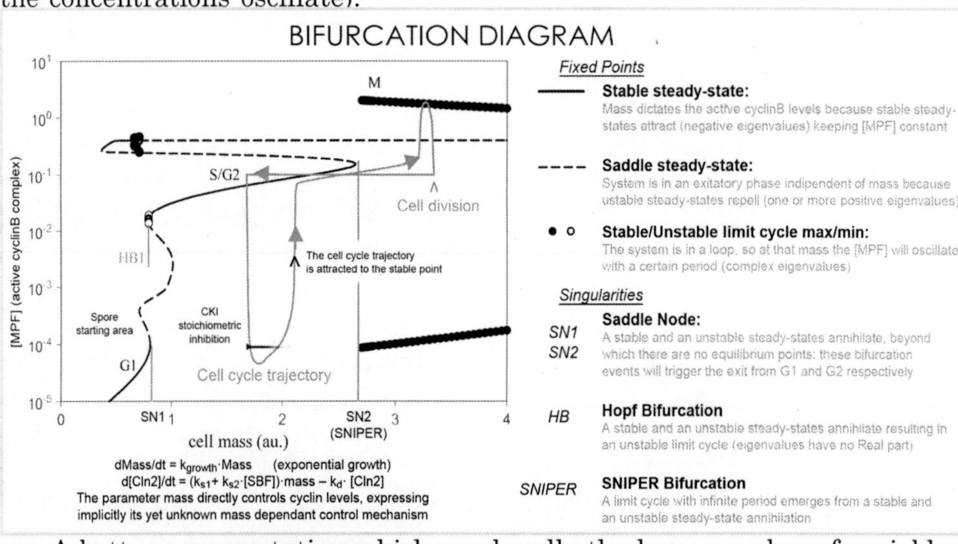

A better representation which can handle the large number of variables and parameters is called a bifurcation diagram (bifurcation theory): the presence of these special steady-state points at certain values of a parameter (e.g. mass) is represented by a point and once the parameter passes a certain value, a qualitative change occurs, called a bifurcation, in which the nature of the space changes, with profound consequences for the protein concentrations: the cell cycle has phases (partially corresponding to G1 and G2) in which mass, via a stable point, controls cyclin levels, and phases (S and M phases) in which the concentrations change independently, but once the phase has changed at a bifurcation event (cell cycle checkpoint), the system cannot go back to the previous levels since at the current mass the vector field is profoundly different and the mass cannot be reversed back through the bifurcation event, making a checkpoint irreversible.

In particular the S and M checkpoints are regulated by means of special bifurcations called a Hopf bifurcation and an infinite period bifurcation.

Molecular Level Simulations

E-Cell Project aims "to make precise whole cell simulation at the molecular level possible".

Projects

Multiple projects are in progress.

- Karyote - Indiana University
- E-Cell Project
- Virtual Cell - University of Connecticut Health Center
- Silicon Cell

Biological Data Visualization

Biology Data Visualization is a branch of bioinformatics concerned with the application of computer graphics, scientific visualization, and information visualization to different areas of the life sciences. This includes visualization of sequences, genomes, alignments, phylogenies, macromolecular structures, systems biology, microscopy, and magnetic resonance imaging data. Software tools used for visualizing biological data range from simple, stand-alone programs to complex, integrated systems.

State-of-the-art and Perspectives

Today we are experiencing a rapid growth in volume and diversity of biological data, presenting an increasing challenge for biologists. A key step in understanding and learning from these data is visualization. Thus, there has been a corresponding increase in the number and diversity of systems for visualizing biological data.

An emerging trend is the blurring of boundaries between the visualization of 3D structures at atomic resolution, visualization of larger complexes by cryo-electron microscopy, and visualization of the location of proteins and complexes within whole cells and tissues.

A second emerging trend is an increase in the availability and importance of time-resolved data from systems biology, electron microscopy and cell and tissue imaging. In contrast, visualization of trajectories has long been a prominent part of molecular dynamics.

Finally, as datasets are increasing in size, complexity, and interconnectness, biological visualization systems are improving in usability, data integration and standardization.

List of Visualization Systems

A large number of software systems are available for visualization biological data. The links below link lists of such systems, grouped by application areas.

Conserved non-coding DNA sequence

A conserved non-coding sequence (CNS) is a DNA sequence of noncoding DNA that is evolutionarily conserved. These sequences are of interest for their potential to regulate gene production.

CNSs in plants and animals are highly associated with transcription factor binding sites and other cis-acting regulatory elements. Conserved non-coding sequences can be important sites of evolutionary divergence as mutations in these regions may alter the regulation of conserved genes, producing species-specific patterns of gene expression.

Spatiotemporal Gene Expression

Spatiotemporal gene expression is the activation of genes within specific tissues of an organism at specific times during development. Gene activation patterns vary widely in complexity. Some are straightforward and static, such as the pattern of tubulin, which is expressed in all cells at all times in life. Some, on the other hand, are extraordinarily intricate and difficult to predict and model, with expression fluctuating wildly from minute to minute or from cell to cell. Spatiotemporal variation plays a key role in generating the diversity of cell types found in developed organisms; since the identity of a cell is specified by the collection of genes actively expressed within that cell, if gene expression was uniform spatially and temporally, there could be at most one kind of cell.

Consider the gene wingless, a member of the wnt family of genes. In the early embryonic development of the model organism Drosophila melanogaster, or fruit fly, wingless is expressed across almost the entire embryo in alternating stripes three cells separated. This pattern is lost by the time the organism develops into a larva, but wingless is still expressed in a variety of tissues such as the wing imaginal discs, patches of tissue that will develop into the adult wings. The spatiotemporal pattern of wingless gene expression is determined by a network of regulatory interactions consisting of the effects of many different genes such as even-skipped and Krüppel.

What causes spatial and temporal differences in the expression of a single gene? Because current expression patterns depend strictly on previous expression patterns, there is a regressive problem of explaining what caused the first differences in gene expression. The process by which uniform gene expression becomes spatially and temporally differential is known as symmetry breaking. For example, in the case of embryonic Drosophila development, the genes nanos and bicoid are asymmetrically expressed in the oocyte because maternal cells deposit messenger RNA (mRNA) for these genes in the poles of the egg before it is laid.

Identifying Spatiotemporal Patterns

One way to identify the expression pattern of a particular gene is to place a reporter gene downstream of its promoter. In this configuration, the promoter gene will cause the reporter gene to be expressed only where and when the gene of interest is expressed. The expression distribution of the reporter gene can be determined by visualizing it. For example, the reporter gene green fluorescent protein can be visualized by stimulating it with blue light and then using a digital camera to record green fluorescent emission.

If the promoter of the gene of interest is unknown, there are several ways to identify its spatiotemporal distribution. Immunohistochemistry involves preparing an antibody with specific affinity for the protein associated with the gene of interest. This distribution of this antibody can then be visualized by a technique such as fluorescent labeling. Immunohistochemistry has the advantages of being methodologically feasible and relatively inexpensive. Its disadvantages include non-specificity of the antibody leading to false positive identification of expression. Poor penetrance of the antibody into the target tissue can lead to false negative results. Furthermore, since immunohistochemistry visualizes the protein generated by the gene, if the protein product diffuses between cells, or has a particularly short or long half-life relative to the mRNA that is used to translate the protein, this can lead to distorted interpretation of which cells are expressing the mRNA.

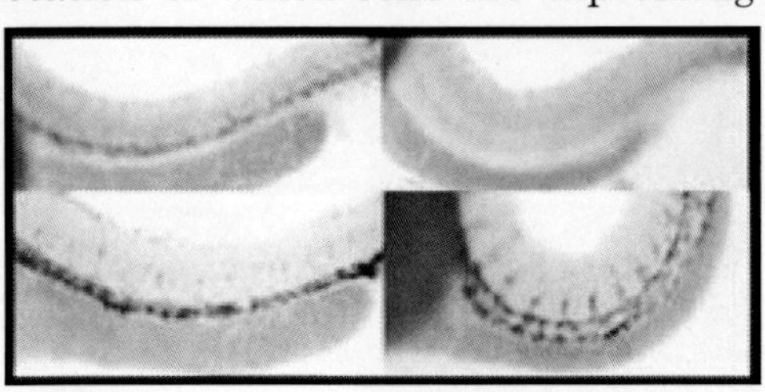

In situ hybridizations of genes expressed in arteries (top) and veins (bottom) in zebrafish. Blue staining indicates presence of the gene mRNAs. Panels on the left are normal animals, while animals on the right are mutated in the Notch gene. Fish lacking Notch have fewer arteries and more veins at this point in developmental time.

In situ hybridization is an alternate method in which a "probe," a synthetic nucleic acid with a sequence complementary to the mRNA of the gene, is added to the tissue. This probe is then chemically tagged so that it can be visualized later. This technique enables visualization specifically of mRNA-producing cells without any of the artifacts associated with immunohistochemistry. However, it is notoriously difficult, and requires knowledge of the sequence of DNA corresponding to the gene of interest.

A method called enhancer-trap screening reveals the diversity of spatiotemporal gene expression patterns possible in an organism. In this technique, DNA that encodes a reporter gene is randomly inserted into the genome. Depending on the gene promoters proximal to the insertion point, the reporter gene will be expressed in particular tissues at particular points in development. While enhancer-trap derived expression patterns do not necessarily reflect the actual patterns of expression of specific genes, they reveal the variety of spatiotemporal patterns that are accessible to evolution.

Reporter genes can be visualized in living organisms, but both immunohistochemistry and in situ hybridization must be performed in fixed tissues. Techniques that require fixation of tissue can only generate a single temporal time point per individual organism. However, using live animals instead of fixed tissue can be crucial in dynamically understanding expression patterns over an individual's lifespan. Either way, variation between individuals can confound the interpretation of temporal expression patterns.

Post-transcriptional Regulation

In eukaryotes, where export of RNA is required before translation is possible, nuclear export is thought to provide additional control over gene expression. All transport in and out of the nucleus is via the nuclear pore and transport is controlled by a wide range of importin and exportin proteins.

Expression of a gene coding for a protein is only possible if the messenger RNA carrying the code survives long enough to be translated. In a typical cell an RNA molecule is only stable if specifically protected from degradation. RNA degradation has particular importance in regulation of expression in eukaryotic cells where mRNA has to travel significant distances before being translated. In eukaryotes RNA is stabilised by certain post-transcriptional modifications, particularly the 5' cap and poly-adenylated tail.

Intentional degradation of mRNA is used not just as a defence mechanism from foreign RNA (normally from viruses) but also as a route of mRNA destabilisation. If an mRNA molecule has a complementary sequence to a small interfering RNA then it is targeted for destruction via the RNA interference pathway.

Translational Regulation

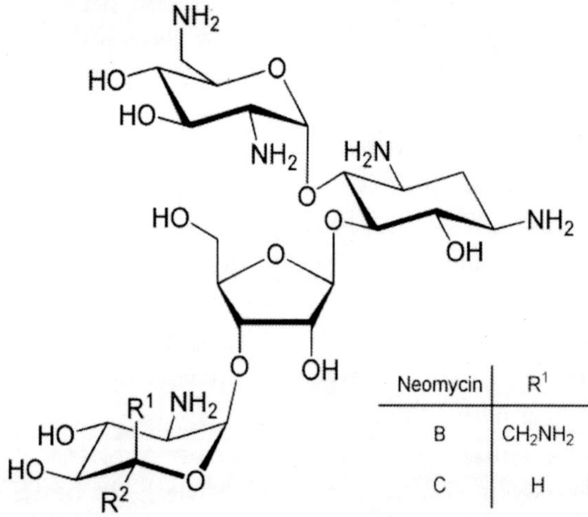

Neomycin	R^1	R^2
B	CH_2NH_2	H
C	H	CH_2NH_2

Figure: *Neomycin is an example of a small molecule which reduces expression of all protein genes inevitably leading to cell death, thus acts as an antibiotic.*

Direct regulation of translation is less prevalent than control of transcription or mRNA stability but is occasionally used. Inhibition of protein translation is a major target for toxins and antibiotics in order to kill a cell by overriding its normal gene expression control. Protein synthesis inhibitors include the antibiotic neomycin and the toxin ricin.

Protein Degradation

Once protein synthesis is complete the level of expression of that protein can be reduced by protein degradation. There are major protein degradation pathways in all prokaryotes and eukaryotes of which the proteasome is a common component. An unneeded or damaged protein is often labelled for degradation by addition of ubiquitin.

Measurement

Measuring gene expression is an important part of many life sciences - the ability to quantify the level at which a particular gene is expressed within a cell, tissue or organism can give a huge amount of information. For example measuring gene expression can:

- Identify viral infection of a cell (viral protein expression)
- Determine an individual's susceptibility to cancer (oncogene expression)
- Find if a bacterium is resistant to penicillin (beta-lactamase expression)

Similarly the analysis of the location of expression protein is a powerful tool and this can be done on an organism or cellular scale. Investigation of localisation is particularly important for study of development in multicellular organisms and as an indicator of protein function in single cells. Ideally measurement of expression is done by detecting the final gene product (for many genes this is the protein) however it is often easier to detect one of the precursors, typically mRNA, and infer gene expression level.

mRNA Quantification

Levels of mRNA can be quantitatively measured by northern blotting which gives size and sequence information about the mRNA molecules. A sample of RNA is separated on an agarose gel and hybridized to a radio-labeled RNA probe that is complementary to the target sequence. The radio-labeled RNA is then detected by an autoradiograph. The main problems with Northern blotting stem from the use of radioactive reagents (which make the procedure time consuming and potentially dangerous) and lower quality quantification than more modern methods (due to the fact that quantification is done by measuring band strength in an image of a gel). Northern blotting is, however, still widely used as the additional mRNA size information allows the discrimination of alternately spliced transcripts.

A more modern low-throughput approach for measuring mRNA abundance is reverse transcription quantitative polymerase chain reaction (RT-PCR followed with qPCR). RT-PCR first generates a DNA template from the mRNA by reverse transcription, which is called cDNA. This cDNA template is then used for qPCR where the change in fluorescence of a probe changes as the DNA amplification process progresses. With a carefully constructed standard curve qPCR can produce an absolute measmsensitive (detection of a single mRNA molecule is possible), but can be expensive due to the fluorescent probes required.

An even more advanced approach is to individually tag single mRNA molecules with fluorescent barcodes (nanostrings), which can be detected one-by-one and counted for direct digital quantification. The advantage of this approach is that it does not rely on analog quantification of fluorescent intensity, which can be problematic due to noise, lack of linearity, and narrow dynamic range. Instead, the technique relies on fluorescence to detect simply the presence of a single mRNA molecule in a binary ("yes" or "no") mode.

Northern blots and RT-qPCR are good for detecting whether a single gene is being expressed, but it quickly becomes impractical if many genes within the sample are being studied. Using DNA microarrays, transcript levels for many genes at once (expression profiling) can be measured. Recent advances in microarray technology allow for the quantification, on a single array, of transcript levels for every known gene in several organism's genomes, including humans. Alternatively "tag based" technologies like Serial analysis of gene expression (SAGE), which can provide a relative measure of the cellular concentration of different mRNAs, can be used. The great advantage of tag-based methods is the "open architecture", allowing for the exact measurement of any transcript, with a known or unknown sequence.

Protein Quantification

For genes encoding proteins the expression level can be directly assessed by a number of means with some clear analogies to the techniques for mRNA quantification.

The most commonly used method is to perform a Western blot against the protein of interest - this gives information on the size of the protein in addition to its identity. A sample (often cellular lysate) is separated on a polyacrylamide gel, transferred to a membrane and then probed with an antibody to the protein of interest. The antibody can either be conjugated to a fluorophore or to horseradish peroxidase for imaging and/or quantification. The gel-based nature of this assay makes quantification less accurate but it has the advantage of being able to identify later modifications to the protein, for example proteolysis or ubiquitination, from changes in size.

Localisation

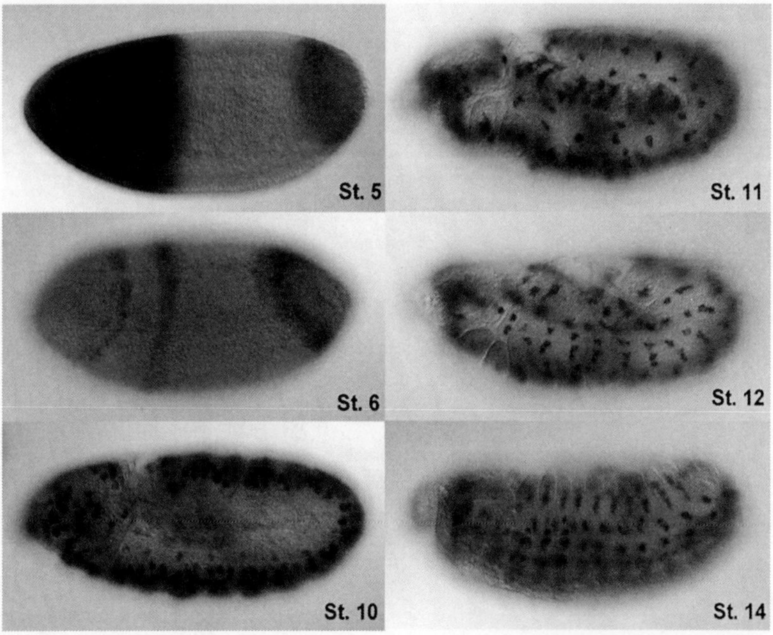

Figure: *In situ-hybridization of Drosophila embryos at different developmental stages for the mRNA responsible for the expression of hunchback. High intensity of blue color marks places with high hunchback mRNA quantity. Analysis of expression is not limited to only quantification; localisation can also be determined. mRNA can be detected with a suitably labelled complementary mRNA strand and protein can be detected via labelled antibodies. The probed sample is then observed by microscopy to identify where the mRNA or protein is.*

By replacing the gene with a new version fused a green fluorescent protein (or similar) marker expression may be directly quantified in live cells. This is done by imaging using a fluorescence microscope. It is very difficult to clone a GFP-fused protein into its native location in the genome without affecting expression levels so this method often cannot be used to measure endogenous gene expression. It is, however, widely used to measure the expression of a gene artificially introduced into the cell, for example via an expression vector. It is important to note that by fusing a target protein to a fluorescent reporter the protein's behavior, including its cellular localization and expression level, can be significantly changed.

The enzyme-linked immunosorbent assay works by using antibodies immobilised on a microtiter plate to capture proteins of interest from samples added to the well. Using a detection antibody conjugated to an enzyme or fluorophore the quantity of bound protein can be accurately measured by fluorometric or colourimetric detection. The detection process is very similar to that of a Western blot, but by avoiding the gel steps moreaccurate quantification can be achieved.

Expression System

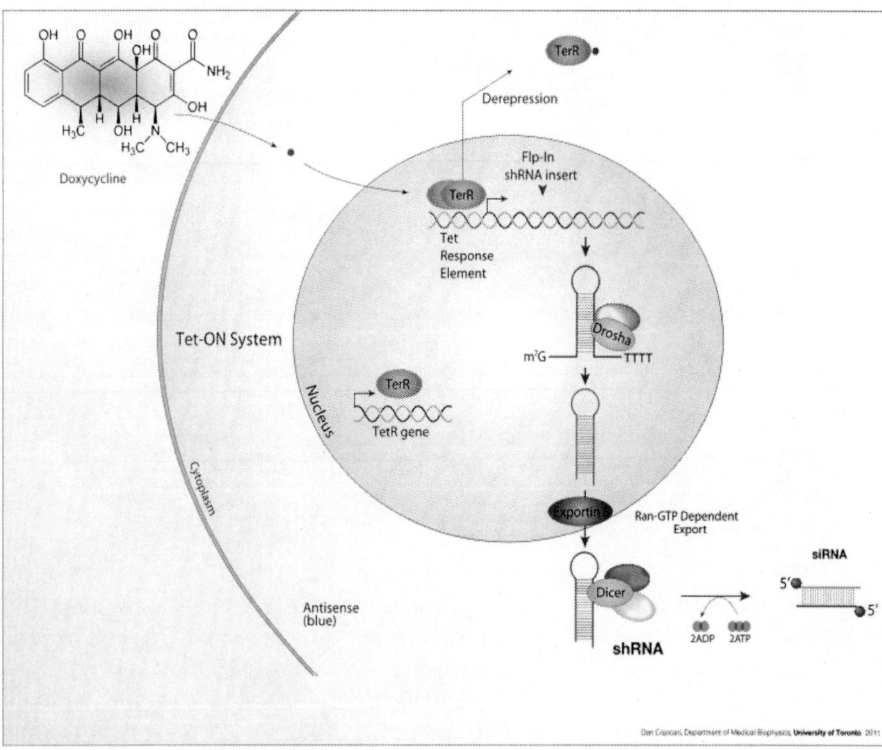

Figure: *Tet-ON inducible shRNA system*

An expression system is a system specifically designed for the production of a gene product of choice. This is normally a protein although may also be RNA, such as tRNA or a ribozyme. An expression system consists of a gene, normally encoded by DNA, and the molecular machinery required to transcribe the DNA into mRNA and translate the mRNA into protein using the reagents provided. In the broadest sense this includes every living cell but the term is more normally used to refer to expression as a laboratory tool. An expression system is therefore often artificial in some manner. Expression systems are, however, a fundamentally natural process. Viruses are an excellent example where they replicate by using the host cell as an expression system for the viral proteins and genome.

Inducible Expression

Doxycycline is also used in "Tet-on" and "Tet-off" tetracycline controlled transcriptional activation to regulate transgene expression in organisms and cell cultures

In Nature

In addition to these biological tools, certain naturally observed configurations of DNA (genes, promoters, enhancers, repressors) and the associated machinery itself are referred to as an expression system. This term is normally used in the case where a gene or set of genes is switched on under well defined conditions. For example the simple repressor switch expression system in Lambda phage and the lac operator system in bacteria. Several natural expression systems are directly used or modified and used for artificial expression systems such as the Tet-on and Tet-off expression system.

5

Chromatin and Chromosomes

Chromatin is the combination of DNA, histone, and other proteins that makes up chromosomes. It is found inside the nuclear envelope of eukaryotic cells. It is divided between heterochromatin (condensed) and euchromatin (extended) forms. The functions of chromatin are to package DNA into a smaller volume to fit in the cell, to strengthen the DNA to allow mitosis and meiosis, and to control gene expression and DNA replication. Changes in chromatin structure are affected by chemical modifications of histone proteins, such as methylation and acetylation, and by other DNA-binding proteins.

Basic Structure

In simple terms, there are three levels of chromatin organization:

- DNA wraps around histone proteins forming nucleosomes - the "beads on a string" structure.
- A 30 nm condensed chromatin fibre consisting of nucleosome arrays in their most compact form.
- Higher-level DNA packaging into the metaphase chromosome.

These structures do not occur in prokaryotic cells. Examples of cells with more extreme packaging are spermatozoa and avian red blood cells.

During spermiogenesis, the spermatid's chromatin is remodelled into a more spaced-packaged, widened, almost crystal-like structure. This process is associated with the cessation of transcription and involves nuclear protein exchange. The histones are mostly displaced, and replaced by protamines (small, arginine-rich proteins).

It should also be noted that, during mitosis, while most of the chromatin is tightly compacted, there are small regions that are not as tightly compacted. These regions often correspond to promoter regions of genes that were active in that cell type prior to entry into cromitosis. The lack of compaction of these regions is called bookmarking, which is an epigenetic mechanism believed to be important for transmitting to daughter cells the "memory" of which genes were active prior to entry into mitosis. This bookmarking mechanism is needed to help transmit this memory because transcription ceases during mitosis.

During Interphase

The structure of chromatin during interphase is optimised to allow easy access of transcription and DNA repair factors to the DNA while compacting the DNA into the nucleus. The structure varies depending on the access required to the DNA. Genes that require regular access by RNA polymerase require the looser structure provided by euchromatin.

Change in Structure

Chromatin undergoes various forms of change in its structure. Histone proteins, the foundation blocks of chromatin, are modified by various post-translational modification to alter DNA packing. Acetylation results in the loosening of chromatin and lends itself to replication and transcription. When certain residues are methylated they hold DNA together strongly and restrict access to various enzymes.

A recent study showed that there is a bivalent structure present in the chromatin: methylated lysine residues at location 4 and 27 on histone 3. It is thought that this may be involved in development; there is more methylation of lysine 27 in embryonic cells than in differentiated cells, whereas lysine 4 methylation positively regulates transcription by recruiting nucleosome remodeling enzymes and histone acetylases.

Nucleocytoplasmic Exchange

Between prokaryotes and eukaryotes resides in the absence or presence of the nuclear envelope. There is the main differences. This structure can be considered as a second permeability barrier that has arisen through evolution to regulate nucleocytoplasmic exchanges and to coordinate gene action with cytoplasmic activity.

One of the most interesting discoveries made with the electron microscope is that the nuclear envelope is a dependency of the cytoplasmic membrane system. This has been verified not only by observing the continuities of the two elements at many points but also by studying these structures in different stages of cellular activity.

The nuclear envelope is a derivative of the endoplasmic reticulum is observed during mitosis. At telophase, cisternae of the endoplasmic reticulum collect around the chromosomes to reform the nuclear envelope. The nuclear envelop consists of two concernic membranes separated by a perinuclear cisterna 10 to 15 nm in width. These membranes have a basic unit structure similar to that of the plasma membrane. The nuclear envelope consists of flattened cisternae of the endoplasmic reticulum having ribosomes only on the outer surface. At certain points the nuclear envelope is interrupted by structures called *pores.*

Around the margins of these nuclear pores both membranes are in continuty. At their nuclear side the pores are generally aligned with channels of nucleoplasm stituated between more condensed lumps of chromatin which attach to the inner membrane. The number of poes varies from 40 to 145 per square micrometer in nuclei of various plants and animals.

At their highest density the pores are packed in hexagonal array with centre-to-centre pore distances of about 150. The nuclear pares are vary large, 50 to 80 nm in diameter. In nuclei of *Mammalia* it has been calculated that nuclear pores account for 5 to 15 per cent of the surface area of the nuclear membrane. In amphibian oöcytes, certain plant cells, and protozoa the surface occupied by the pores may be as high as 20 to 36 per cent. Nuclear pores, however, are not wide-open channels, the electron microscope has revealed that they are occluded by an electron-dense materia. The pores are enclosed by circular structures called *annuli.*

The pores and annuli are together designated the *pore complex.* Negative staining techniques have demonstrated that the annuli are octagonal in shape and are about 60 nm in diameter. Each annulus consists of eight *granules* about 15 nm, which are present on both the nuclear and cytoplasmic surfaces and a less defined, amorphous *annular material* in the opening itself.

This material is digested by trypsin and remains unaffected when exposed to ribonuclease and deoxyribonuclease. These findings suggest that the annular material is protein in nature. The pore complex is apparently a rather rigid structure present in a fixed number according to cell type. In certain physiologic stages, however, they may change in number.

For example, they are reduced in number in maturing erythroblasts and in spermatids, and this correlates with the low transcriptional activity of these cells. In the frog *Xenopus laevis,* oöcytes (which are very active in transcription) have 60 pores/μm^2 (and up to 30 million pore complexes per nucleus), whereas mature erythrocytes (inactive in transcription) have only about 3 pores/μm^2 (and a total of only 150 to 300 pores per nucleus. The nuclear envelope has been isolated from a variety of cells and analysed biochemically. The main conclusion to be derived from these studies is that the composition of the nuclear membrane is very similar to that of the rough endoplasmic reticulum.

This emphasizes the relationship between both membrane systems, which can also be visualized by their physical continuity and by the presence of ribosomes attached to the outer nuclear membrane.

The developing oöcytes and spermatocytes, as well as in certain embroynic and tumor cells, the so-called *annulated lamellae* may be found in the cytoplasm. These membranous structures appear as stacks of cisternae or flattened sacs with an intrasaccular space 10 to 20 nm wide. At regular intervals these sacs are traversed by pore complexes that are similar in morphology and structure to those found in the nuclear envelope. These are closely spaced and may occupy up to 50 per cent of the surface of the membrane.

The lamellae have been interpreted as budding off the nuclear envelope. More precise work has demonstrated that first there is a budding of the outer membrane of the nuclear envelope into flattened sacs; later the pore complexes appear. They are usually associated with rapidly proliferating cells. The material exchanged between nucleus and cytoplasm must traverse the nuclear pore complexes. This exchange is very selective and allows passage of only certain molecules, of either low or very high molecular weight.

The permeability of the nuclear membrane with respect to small ions is reflected in some of the electrochemical properties of this structure, which can be investigated with fine microelectrodes. When giant cells from the salivary gland of *Drosophila* are penetrated with a microelectrode, there is an abrupt change in potential at the plasma membrane (— 12 mV); then, as the microelectrode enters the nucleus, there is another drop in negative potential at the nuclear membrane (—13 mV).

These results suggest that the nuclear envelope is a diffusion barrier for ions as small as K^+, Na^+, or Cl^-. On the other hand, very large structures such as ribosomal subunits, which are assembled in the nucleolus, are able to leave the nucleus.

All RNAs can exit freely; heterogeneous nuclear RNA (hnRNA, a large precursor of mRNA which contains intervening sequences is never found in the cytoplasm, and the mRNA sequences can traverse the pores only after extensive processing has occurred within the nucleus. Xenopus oöcytes are particularly useful for studying nucleocytoplasmic exchanges because these cells are very large (up to 1.2 mm diameter), and substances can be easily microinjected into the cytoplasm. One can then find out whether or not these substances entered the nucleus by manually isolating the giant oöcyte nucleus (also called the germinal vesicle), which measures up to 0.4 mm in diameter. Studies involving the microinjection of dextrans, colloidal gold particles, and labeled proteins sugest that the oöcyte nuclear membrane has functional pors with a radius of about 4.5 nm. It was found that small proteins (such as lysozyme, MW 14,000; and ovalbumin, MW 44,000), were able to enter the nucleus, that bovine serum albumin (MW 62,500) entered extremely slowly, and that larger proteins did not enter at all. This upper limit for diffusion applies only to proteins that are not normally nuclear proteins for, as we will see in the following section, nuclearspecific proteins have differently.

The cell nucleus contains a specific subset of proteins. For example, RNA polymerase, DNA polymerase, and histones are located specifically inside the nucleus. However, all these proteins are synthesized in the cytoplasm and must be subsequently transported into the nucleus. Here we will discuss experiments which show that nuclear proteins contain, within the structure of the nature protein, a signal that enables them to accumulate in the nucleus after microinjection into the cytoplasm.

This is observed because when labeled proteins, such as histones or a preparation of nucleoplasmic proteins, are microinjected into the cytoplasm of an oöcyte, they are not only able to enter the nucleus but also to *accumulate* in it.

When the nuclear proteins of normal oöcytes are examined by two dimensional gel electrophoresis, two types of proteins can be distinguished:

1. Nucleus specific proteins,
2. Proteins common to *both* nucleus and cytoplasm.

Actin is the most conspicuous member of the "common-to-both" proteins, and it is one of the major constituents of the oöcyte nucleoplasm.

A labeled preparation of *Xenopus* hucleoplasmic proteins (which contained actin and nuclear-specific proteins. After 24 hours the oöcytes were manually separated into nucleus and cytoplasm, and the proteins were analysed by two-dimensional gels.

It was found that nucleus-specific proteins were greatly concentrated in the nucleus (100-fold in the case of N1), while microinjected actin was found both in nucleus and cytoplasm. This is as expected, since actin is normally present in both cell compartments. The nuclearspecific proteins here in their molecular structures some type of signal that enables them to accumulate selectively in the nucleus. This signal must be absent in proteins which do not accumulate in the nucleus, such as actin.

Some of the nuclear proteins are quite large (N1 has a MW of 120,000 daltons) but are still able to re-enter the nucleus rapidly, it therefore seems that when proteins are nucleus-specific, factors other than size determine nuclear pore permeability. Cells have the capacity to place specific proteins into the corresponding cell compartments, as occurs with the mitochondria or chloroplast proteins that are synthesized in the cytoplasm, or with proteins that are exported from the cell. The mechanism by which proteins are secreted into the endoplasmic reticulum is very different from the one used for the accumulation of proteins inside the nucleus.

The "signal peptide" enables the secreted proteins to traverse the endoplasmic reticulum membrane only as they are being synthesized, and it is rapidly removed thereafter; the nuclear proteins, however, accumulate in the nucleus independently of protein synthesis, and this is not a transient property but is intrinsic to the structure of the mature protein.

Genetic Constituent

Cell have the genetic constituent *i.e.*, DNA, carrying information in a coded from from cell to cell and from organism to organism. Within cells, DNA is not free but is complexed with proteins in a structure called chromatin.

The domonstration that nuclei contain amount of DNA which is a landmark in cell biology because it suggested that DNA was the genetic material and also that genetic information is not lost during the differentiation of the various somatic tissues.

This was achieved using cytophotometric methods. The DNA in nuclei was stained using the Feulgen reaction, and the amount of stain was estimated by microcytophotometry. Cytophotometry has produced accurate measurements in a large variety of cell types within the same species, showing that the DNA content is practically constant in the different tissues. This is certainly related to the number of chromosome sets (*n*), which is 2n (diploid) in most somatic cells. However, certain tissues, like liver, have some cells that become polyploid (4n or 8n), and their nuclei have a correspondingly higher DNA content.

The other example which is provided by the study of spermatogenesis. Before meiosis occurs, there are two classes of cells (spermatogonia) having different DNA contents (2n and 4n).

The early primary spematocyte (meiotic prophase) has a tetraploid amount of DNA. After the first meiotic division, the secondary spermatocyte contains half the DNA content, corresponding to a diploid condition. Finally, the second meiotic division results in four spermatids; these have the DNA of only one chromosomal set (haploid cell). Similar results have been observed in oögenesis.

Each species has a characteristic amount of DNA. Eukaryotes vary greatly in DNA content but always contain much more DNA than prokaryots. Lower eukaryotes in general have less DNA, such as the nematode *C. elegans,* which has only 20 times more DNA than *Escherichia coli,* or the fruit fly *Drosophila melanogaster*, which has 40 times more. Vertebrates have greater DNA content, in general about 700 times more than *E. coli.* The highest DNA content is that of the salamander *Amphiuma,* Which can contain 168 picograms of DNA in a single nucleus, *i.e.*, nearly 40,000 times more than *E. coli.* It is considered that one pictogram of DNA is equivalent to 31 cm of DNA, it is possible to calculate that there are about 174 cm of DNA in human diploid cells, 37 meters in *Trillium* and 97 meters in polytenic chromosomes of *Drosophila.*

The DNA content in the 46 human chromosomes has been estimated by cytophotometry, and from these measurements it appears that the DNA content is proportional to the size of the chromosome. The largest chromosome, which is 10 mm long, should accommodate about 7.2 cm of DNA in a tightly packed form; *i.e.*, the DNA is compacted about 7000-fold in a metaphase chromosome! In this chapter we will analyse how this folding is achieved. During interphase, chromatin can be present in a condensed state (heterochromatin) or in a more dispersed form (euchromatin). The heterochromatic regions tend to form aggregates, sometimes called chromocentres.

Thin-section electron microscopy contributed little to our knowledge of the organization of chromatin in the interphase nucleus, in contrast to the results obtained by electron microscopic analysis of *spread* preparations. Asure have already shown in previous chapter, the distribution of chromatin in a nucleus stained with uranyl acetate. Both RNA and DNA are heavily stained, but DNA is most conspicuous. The placement of the chromatin in clumps near the nuclear envelope and the interchromatin channels leading to the nuclear pores are clearly visible. The mass of the nucleolus is of a lesser density and is surrounded by an associated chromatin of a greater density; the nucleoplasm is light and reveals fine granules.

The nucleus has also been stained with uranyl acetate, but it was then treated with EDTA (ethylenedi-aminetetraacetic acid) which preferentially removes the uranyl ions from the DNA. EDTA is a chelating agent which binds to the phosphate groups of DNA, thus displacing the UO_2^{2+} ions. In this way, the RNA-containing structures of the nucleus and the ribosomes in the cytoplasm become more conspicuous. DNA structure shows a high electron density.

Chromatin can be isolated biochemically by purifying nuclei and then lysing them in hypotonic solutions. When prepared in this way, chromatin appears as a viscous, gelatinous substance which contains DNA, RNA, basic proteins called histones, and nonhistone (more acidic) proteins. The content of RNA and nonhistone protein is variable between different chromatin preparations, but histone and DNA are always present in a fixed ratio of about one to one. The nonhistone proteins are very heterogeneous; they vary in different tissues and include RNA and DNA polymerases, among other enzymes, and putative regulatory proteins. On the other hand, there are only five types of histones, each one present in large amount and this suggested that they could have a structural role. Some of the properties of histones. Histones are small proteins which are basic because they have a high content (10 to 20 per cent) of the basic amino acids arginine and lysine. Being basic, histones bind tightly to DNA, which is acid. The four main histones, H2A, H2B, H3, and H4, are very similar in different species. For example, the sequence of histone H3 from the rat differs only in two-amino acids from that of peas. These four histones are present in equimolar amounts (two of each every 200 base pairs of DNA).

Histone H1 is not conserved between species and has tissue-specific forms; is present only once per 200 base pairs of DNA; is rather loosely associated with chromatin (if can be eluted from DNA by adding low concentrations of salt); and is not a component of the nucleosome but is involved in the maintenance of a higher-order folding of chromatin. Prokaryotes do not have histones, but a basic protein that might serve the same function has been isolated from *E. coli*.

When eukaryotic nuclei were *spread* on EM specimen holders (instead of prepared as thin sections), it was found that chromatin has a repeating structure of beads about 10 nm in diameter connected by a string of DNA. Most, if not all, of the DNA is present in this form. The beads-on-a-string appearance is not the true structure of chromatin but is an artifact arising from the way in which the sample is prepared, which results in a loss of histone H1. With more mild treatments (which do not remove H1), the beads are not stretched but can be observed as a 10 nm fibre, with the beads touching each other. The 10nm fibre therefore represents the first level of organization of chromatin within cells.

The Nucleosome

Studies of digestion of chromatin by nucleases 22 showed that the DNA was cut into multiples of a unit size, which was later found to be about 200 base pairs of DNA in length; *i.e.*, the DNA was cut at intervals of 200, 600, 800 base pairs and so forth. Other studies showed that histones H3 and H4 tend to associate in solution forming tetramers consisting of two of each. Since the four histones are in equimolar amounts (two of each per 200 base pairs of DNA), in 1974 R. Kornberg (son of Arthur Kornberg, the discoverer of DNA polymerase) proposed a model for the nucleosome, in which the four histones H2A, H2B, H3, and H4 are arranged in octamers containing two of each of them, every 200 base pairs of DNA.

The DNA is cut into discrete fragments by micrococcal nuclease, because the linker regions between two neighboring nucleosomes are more exposed to the enzyme. Very extensive digestion leads to the degradation of the linker DNA and generates a nucleosome *core particle*, which contains 140 base pairs of DNA and lacks the 60 base pairs of the linker DNA. Core particles lack histone H1, indicating that H1 is normally associated with the linker DNA. Core nucleosomes have been crystallized, and x-ray diffraction and EM studies have shown that the nucleosome is a flat disk about 11 nm in diameter and 5.7 nm in height. The DNA is on the outside and turns twice around each histone octamer.

The length of DNA per nucleosome can vary in different tissues. For example, chicken erythrocytes have a repeat length of 210 base pairs, and chicken liver has about 195 base pairs per repeat. The reason for this is not clear, but it seems to be related to the type of histone H1 in each tissue. The variability is restricted to the length of the linker segment, and all tissues have a constant amount of DNA 140 base pairs) in the nucleosome core. When nucleosomes are in close apposition in the 10 nm filaments, the packing of DNA is about five-to seven-fold, *i.e.*, five to seven times more compact than free DNA. However, this is still 1000-fold lower than the packing ratio of DNA in metaphase chromosomes, which is achieved by further folding the 10 nm fibre. Underlectron microscopic studies of chromatin fibres in interphase nuclei and mitotic chromosomes; have revealed a "*thick fibre*" of a diameter that varies from 20 to 30 nm and which probably represents the structure of inactive chromatin. You can seek that the 20 to 30 nm fibre consists of closely packed nucleosomes. The 20 to 30 nm fibre probably arises from the folding of the nucleosome chain into a solenoidal structure having about six nucleosomes per turn. The histone H1 is located in the central "hole" of the solenoid, and the whole structure is stabilized by interactions between different H1 molecules. Solenoidal structures can be formed in vitro by adding Mg^{2+} to 10 nm fibre preparations, but they can not be formed when histone H1-depleted chromatin is treated in the same way.

The DNA of a 20 to 30 nm solenoid has a packing that is about 40-fold. However, the DNA of a metaphase chromosome is packed between 5000 and 10,000 times; *i.e.*, the 20 to 30 nm fibre must be further folded more than 100-fold. At present, we know very little of how this feat is achieved.

The Chromosomes

A chromosome is an organized structure of DNA and protein that is found in cells. It is a single piece of coiled DNA containing many genes, regulatory elements and other nucleotide sequences. Chromosomes also contain DNA-bound proteins, which serve to package the DNA and control its functions. Chromosomes vary widely between different organisms. The DNA molecule may be circular or linear, and can be composed of 10,000 to 1,000,000,000 nucleotides in a long chain. Typically eukaryotic cells (cells with nuclei) have large linear chromosomes and prokaryotic cells (cells without defined nuclei) have smaller circular chromosomes, although there are many exceptions to this rule. Furthermore, cells may contain more than one type of chromosome; for example, mitochondria in most eukaryotes and chloroplasts in plants have their own small chromosomes.

In eukaryotes, nuclear chromosomes are packaged by proteins into a condensed structure called chromatin. This allows the very long DNA molecules to fit into the cell nucleus. The structure of chromosomes and chromatin varies through the cell cycle. Chromosomes are the essential unit for cellular division and must be replicated, divided, and passed successfully to their daughter cells so as to ensure the genetic diversity and survival of their progeny. Chromosomes may exist as either duplicated or unduplicated—unduplicated chromosomes are single linear strands, whereas duplicated chromosomes (copied during synthesis phase) contain two copies joined by a centromere.

Compaction of the duplicated chromosomes during mitosis and meiosis results in the classic four-arm structure (pictured to the right). Chromosomal recombination plays a vital role in genetic diversity. If these structures are manipulated incorrectly, through processes known as chromosomal instability and translocation, the cell may undergo mitotic catastrophe and die, or it may unexpectedly evade apoptosis leading to the progression of cancer.

In practice "chromosome" is a rather loosely defined term. In prokaryotes and viruses, the term genophore is more appropriate when no chromatin is present. However, a large body of work uses the term chromosome regardless of chromatin content. In prokaryotes DNA is usually arranged as a circle, which is tightly coiled in on itself, sometimes accompanied by one or more smaller, circular DNA molecules called plasmids. These small circular genomes are also found in mitochondria and chloroplasts, reflecting their bacterial origins. The simplest genophores are found in viruses: these DNA or RNA molecules are short linear or circular genophores that often lack structural proteins.

During cell division chromatin becomes condensed into the chromosomes. Since chromosomes can be clearly visualized with the optical microscope, they were the subject of much study soon after their discovery in 1876, and by 1910 it became clear that genetic phenomena could be explained in terms of chromosomes behaviour.

For this very same reason, the study of chromosomes remains of the utmost importance even today. In this section we will analyse the general morphology and composition of chromosomes.

Chromosomes can be studied in tissue sections but they are best visualized in *squash* preparations. Fragments of tissues are stained with basic dyes (*e.g.*, orcein or Giemsa) and then squashed between slide and coverslip by gentle pressure. Sometimes hypotonic solutions are used prior to squashing to produce swelling of the nucleus and a better separation of the individual chromosomes. The morphology of the chromosomes is best studied during metaphase and anaphase, which are the periods of maximal contraction.

As per classification, there are four types according to their shape, which, in turn, is determined by the position of the centromere (point of attachment to the mitotic spindle). *Telocentric* chromosomes have the centromere located on one end; *acrocentric* chromosomes have a very small or even imperceptible short arm; *submetacentric* chromosomes have arms of unequal length; and *metacentric* chromosomes have equal or almost equal arms. During anaphase movements the chromosomes bend at the centromere, so that metacentric chromosomes are v-shaped, and acrocentric chromosomes are rod-shaped. This angular deviation can also be observed at metaphase (provided that the cells are not treated with colchicine).

As per classification, cytogeneticists produced many names to describe the various components that can be visualized in mitotic chromosomes. The student will probably find this terminology initially confusing, but it is important to become familiar with these names, for they not only define particular structures but also specific properties of the chromosomes.

Chromomeres

Chromomeres are bead like accumulations of chromatin material that are sometimes visible along the chromonema. Chromomeres are especially obvious in polytene chromosomes, where they become aligned side by side, constituting the chromosome bands. These tightly folded regions of the chromonema are of considerable interest, because they are thought to correspond to the units of genetic function in the chromosome. At metaphase the chromosome is tightly coiled, and the chromomeres are no longer visible.

Centromere

The centromere lies within a thinner segment of the chromosome, the *primary constriction*. The regions flanking the centromere frequently contain highly repetitive DNA and may stain more intensely with basic dyes (heterochromatin).

Kinetochore

The centromere surrounded by kinetcore shaped. Unlike the other chromosome components, the kinetochore can be studied best in thin sections with the electron microscope, which reveals a trilaminar structure. The kinetochore has an dense outer layer, middle layer of low density, and a dense inner layer tightly attached to the centromere. The function of the kinetochore is to provide a centre of assembly for microtubules; when isolated metaphase chromosomes are incubated in a solution of tubulin with an appropriate buffer, microtubules are reassembled at the kinetochores of the chromosome.

Chromatid

At metaphase each chromosome consists of two symmetrical structures; the *chromatids;*, each one of which contains a single DNA molecule. The chromatids are attached to each other only by the centromere and become separated at the start of anaphase, when the sister chromatids migrate to opposite poles. Therefore, *anaphase chromosomes have only one chromatid, while metaphase chromosomes have two.*

Chromonemata

During prophase and some times during interphase the chromosomal material becomes visible as very thin filaments, which are called chromonemata and which represent chromatids in early stages of condensation. "Chromatid" and "chromonema," therefore, are two names for the same structure: a single linear DNA molecule with its associated proteins. The vast majority of chromosomes have only one centromere. Some species have diffuse centromeres, with microtubules attached along the length of the chromosome (*holocentric chromosomes*).

In some chromosomal abnormalities (induced for example by x-rays), chromosomes may break and fuse with other ones producing chromosomes without centromeres (*acentric*) or with two centromeres (*dicentric*). Both types of aberration are unstable; one because it cannot attach to the mitotic spindle and remains in the cytoplasm, the other because the two centromeres tend to migrate to opposite poles, thus leading to chromosomal fragmentation.

Telomere

The term 'Telomere applies to the tips of the chromosomes, which have distinct cytological properties. When chromosomes are broken by x-rays, the free ends without telomeres become "sticky" and fuse with other broken chromosomes; they will not, however, fuse with a normal telomere. Stable deletions in chromosomes are always interstitial and preserve the original telomere. In addition to promoting chromosomal stability, in some species telomeres mediate transient end-to-end association of chromosomes. A telomeric DNA sequence which is present at the tips of all the *Drosophila melanogaster* chromosomes has been isolated. Telomeres contain the ends of the long linear DNA molecule contained in each chromatid.

Secondary Constrictions

These are the other morphologic characteristics. Constant in their position and extent, these constrictions are useful in identifying particular chromosomes in a set. Secondary constrictions are distinguished from the primary constriction by the absence of marked angular deviations of the chromosomal segments.

Nucleolar Organizers

These areas are certain secondary constrictions which contain the gene coding for 18S and 28S ribosomal. The and which induce the formation of nucleoli. The secondary constriction arises because the rRNA genes are transcribed arises because the rRNA genes are transcribed very actively, interfering with chromosomal condensation. In man, the nucleolar organizers are located in the secondary constrictions of chromosomes, 13, 14, 15, 21, and 22, all of which are acrocentrics and have satellites.

Satellites

Another morphologic element present in certain chromosomes is the satellite. This is a rounded body separated from the rest of the chromosomes by a secondary constriction. The satellite and the constriction are constant in shape and size for each particular chromosomes. The term Karyotype is concerned with the whole group of characteristics that allows the identification of a particular chromosomal set; *i.e.*, the number of chromosomes, relative size, position of the centromere, length of the arms, secondary constrictions, satellites, and so forth.

The karyotype is characteristic of an individual, species, genus, or larger grouping, and may be represented by a diagram in which the pairs of homologues are ordered in a series of decreasing size. Some species may have special characteristic; for example, the mouse has acrocentric chromosomes, many amphibia have only metacentric chromosomes, and plants frequently have heterochromatic regions at the telomeres.

Some species are particularly favorable for cytogenetic analysis: Salamanders and grasshoppers have high DNA content and very large chromosomes which produce very beautiful meiotic preparations; among plants, broad beans (*Vicia fava*) and onions (*Allium cepa*) have particularly large chromosomes which produce nice preparations of mitotic cells from root tips. Each chromatid represents a single linear DNA molecule with outs associated proteins. This concept, sometimes called the *unineme theory*, is supported by several lines of evidence semiconservative replication of a single DNA molecule. When cells are labeled with ^{3}H-thymidine or with bromode-oxyuridine (an analogue that becomes incorporated into DNA and prevents the staining of the chromatid with certain dyes) and then grown for two cycles in an unlabeled medium, only one of the granddaughter chromatids is labeled, as expected for a single DNA molecule.

Furthermore, exchanges between sister chromatids can be induced experimentally, and in this case every change in chromatid is accompanied by a reciprocal change in the other, a fact which is also consistent with the idea that there is only one DNA molecule with the finding that G_1 chromosomes (which have not yet replicated their DNA) have only one chromatid, and that G^2 chromosomes have two. The unineme organization of the chromatid is supported directly by the isolation of DNA molecules that are long enough to contain all the DNA from a single chromatid. The length of these giant molecules can be estimated by viscoelastic measurements.

The two sister chromatids are mirror images one of the other. The morphologic characteristics of one chromatid, such as chromomeres, secondary constrictions, and satellites, always have their counterparts in the other. The reason for this symmetry is that sister chromatids contain identical DNA molecules, and the morphologic features of chromosomes are ultimately determined by the DNA sequence they contain.

The condensation and decondensation cycles of the chromatin fibres during mitosis and meiosis would be greatly faciliated if each chromosome were maintained in a definite space within the nuclear cavity during interphase, so as to avoid entanglements of the chromatin fibres. Some evidence exists in plant cells to indicate that the chromosomes retain during interphase the same spatial order they had at telophase. This was shown to be true in onion chromosomes, which have characteristic heterochromatic regions, both at the centromeres and at the telomeres.

Throughout interphase the centromeres remain on one side of the nucleus, whereas the telomeres are on the opposite side, in the same orientation that occurs during telophase. It is even possible that chromosomes may be arranged in a fixed pattern with respect to each other as is suggested by the classical work of D. P. Costello, who analysed the first embryonic division in a turbellarian that has chromosomes that can be recognized by particular morphologic features. The chromosomes were arranged in a precise order during metaphase. The probability that such an arrangement would arise at random is less than 1 in 2.

Condensation and decondensation cycles. The interphase chromatin becomes condensed at mitosis, and the chromatids become increasingly shorter from prophase to metaphase, and they become decondensed at telophase. In some favorable organisms, such as the plants *Trillium* and *Tradescantia*, the optical microscope shows that mitotic condesation involves a helical coiling of the chromatids. In other species, including man, the coils become apparent only after certain pretreatments that partially open the tight helix.

Here is showing that mitotic chromosomes consist of chromatin fibres (20 to 30) nm in diameter. In these whole mount chromosomal preparations the chromosomes are dispersed on an air-water interface and then dehydrated by the critical-point drying method of Anderson. In this process the water content is replaced by liquid CO_2 in a pressure chamber, and then the specimen is heated above the *critical temperature* (*i.e.*, 31°C), at which liquid CO_2 is converted suddenly into gas.

The disorderly folding of the 20 to 30 nm fibres that can be seen in these preparations is an artifact due to the spreading on water. In the living cell, the chromatin fibres must be much more ordered, since an orderly structure can be packed into a much smaller space than a disorganized one. As discussed earlier, a metaphase chromosome is 5000 to 10,000 times shorter than the equivalent amount of free DNA. By comparison, the DNA in the head of bacteriophage T_4 is packed at a ratio of 520/1.

The DNA in mitotic chromosomes is contained in nucleosome chains. which are in turn coiled into 20 to 30 nm chromatin fibres, this represents a packing ratio of 40/1. The way in which the higher levels of packaging are achieved is not known, but some ideas about this are emerging from the study of histone-depleted chromosomes.

Structural nonhistone proteins could be involved in the higher-order coiling of the 20 to 30 nm chromatin fibres. It is possible to remove the histones from metaphase chromosomes (by adding the polyanion dextran sulfate, which competes with the DNA for the basic histones).

Histone-depleted chromosomes have a central core or *scaffold*, surrounded by a halo made of loops of DNA. The scaffold is made of nonhistone proteins and retains the general shape of the metaphase chromosome. Each chromosome has two scaffolds, one for each chromatid, and they are connected together at the centromere region. When the histones are removed, the DNA, which was packed about 40-fold in the chromatin fibres, becomes extended and produces loops with an average length of 25 μm (75,000 base pairs).

A in each loop the DNA exits from the scaffold and returns to an adjacent point. On the basis of these observations a model of chromosome structure has been proposed in which the DNA is arranged in loops anchored to the nonhistone scaffold.

Since the lateral loops have 25 μm, of DNA, after contracting 40-fold in the 20 to 30 nm fibre, they would be only about 0.6 μm long, a length consistent with the diameter of metaphase chromosomes (1 μm) and with micrographs such as the one.

Other work, using very different methodology, has suggested that it is possible that even in interphase nuclei the chromatin might be organized in "domains" or loops containing about 80,000 base pairs of DNA. This raises the possibility that the DNA domains or loops might persist throughout the cell cycle. More work will be required before we understand the molecular organization of chromosomes.

Heterochro Matin Region

In 1928 Heitz defined *heterochromatin* as those regions of the chromosome that remain condensed during interphase and early prophase and form the so-called *euchromatin (Gr., eu,* true). Heitz followed cells throughout the cell cycle and found chromosomal segments that do not decondense.

The heterochromatic segments tend to show preterential localizations in the centromeres of most plants and animals, at the telomeres (especially in plants), or intercalated in the chromosomal arms (frequently adjacent to the nucleolar organizers). In other cases, whole chromosomes become heterochromatic.

The heterochromatic regions can be visulized in condensed chromosomes as regions that stain more strongly or weakly than the euchromatic regions, showing what is called a *positive* or a *negative heteropyknosis* of the chromosomes (Gr., *hetero +pyknosis,* different staining).It is thought that in heterochromatin the DNA remains tightly packed in the 20 to 30 nm fibre, which probably represents the configuration of transcriptionally inactive chromatin.

Heterochromatin

Two types of heterochromatin are generally recognized constitutive heterochromatin which is permanently condensed in all types, of cells, and *facultative* heterochromatin, which is condensed only in certain cell types or at special stages of development. Frequently in *facultative* heterochromatin one chromosome of the pair be comes either totally or partially heterochromatic. The best known case is that of the X chromosomes in the mammalian female, one of which is active and remains euchromatic, whereas the other is inactive and forms the sex chromatin, or...body, at interphase

Constitutive heterochromatin is the most common type of heterochromatin. Most chromosomes contain large blocks of heterochromatin at the centromeres, frequently comprising 5 to 10 per cent of the total chromosomal DNA and, in some extreme cases, such as in *Drosophila virilis,* it can constitute up to 40 per cent of the chromosome. This type of heterochromatin contains highly repeated DNA sequences, which might have a structural role in chromosomes.

A property common to all types of heterochromatin is *late replication.* When cells are given a brief pulse of ^{3}H-thymidine during the late S phase, the label is incorporated only into the heterochromatic segments, indicating that they replicate after the bulk of the DNA. It is generally agreed that condensed chromatin is inactive in RNA synthesis. For example, in the next chapter we will see that condensed mitotic chromosomes do not synthesize RNA. There is good genetic evidence indicating that genes contained in heterochromatic segments are not expressed. Here, we will examine three of these lines of evidence.

Some cats have a striking spotted coloring of the coat. Because of the peculiar coloured patches of the coat, these cats are known as *tortoiseshells.* Tortoiseshell cats are always female, never male. The reason for this become clear when Mary Lyon put forward the hypothesis that the patchy pigmentation is produced by a gene contained in the X chromosome, which becomes heterochromatic and inactive in some groups of cells put not in others.

One of the X chromosomes in mammalian females becomes inactivated early in embryogenesis on a random basis, so that adult animals are mosaics in which 50 per cent of the cells have the paternal X chromosome euchromatic and active, and the other 50 per cent have the maternal X chromosome in an active state. Another striking instance of facultative heterochromatin is found in the mealy bug (*planococcus citri)*, an insect in which the males have a haploid set of chromosomes that are entirely heterochromatic. The heterochromatic set is the paternal one.

The genes in these paternal chromosomes are switched off, as demonstrated by the fact that it is not possible to induce lethal mutations in them, even by high does of radiation, whereas dominant lethals can be readily obtained by irradiation of the euchromatic set. In female mealy bugs both sets of chromosomes are euchromatic and both are genetically active; *i.e.*, lethal mutations can be induced in both the paternal and the maternal set. Genes can also be inactivated by translocation into heterochromatic regions of the chromosomes. This was first shown in *Drosophila melanogaster,* in which a gene coding for eye colour (w) can be inactivated by translocating it close to the centromeric heterochromatin.

The biological significance of the switching off of genes in condensed chromatin might be more universal than indicated in the limited examples discussed here. Multi-cellular organisms have many specialized tissues, and in each one of them a particular set of genes is active and others are inactive. For example, globin genes are expressed in red blood cells but are not expressed in other tissues. We do not know how entire sets of genes are switched off, but many believe that variable degrees of chromatin condensation could be involved in the maintenance of the differentiated state.

Constitutive heterochromatin is related to repetitive DNA. In many cells, it is possible to demonstrate the presence of special DNA fractions using gradients of cesium chloride.

By using the satellite DNA as the template for in vitro synthesis, it is possible to obtain a ^{3}H-RNA can be hybridized to the satellite DNA not only in the test tube but also directly on a slide with the cells that contain this type of DNA. The hybridization, also called *in situ hybridization,* was detected by autoradiography and localized specifically at the centromeric regions of all the chromosomes. This experiment demonstrates that the satellite DNA is present in the centromeric heterochromatin.

The Nucleolus

The presence of nucleoli, which appear as dense granules within the nucleus, was first described by Fontana in 1781. By the end of the 19 th century a relationship between the size of the nucleolus and the synthetic activity of the cell was postulated. It was found that nucleoli were small or absent in cells exhibiting little protein synthesis (sperm cells, blastomeres, muscle cells, and so forth), whereas they were large in oöcytes, neurons, and secretory cells — those in which protein synthesis is a prominent feature. In the living cell, nucleoli are highly refringent bodies. This is the result of a large concentration of solid material, which may constitute 40 to 85 per cent of the dry mass. The light microscope generally reveals the nucleolus as structurally homogeneous, although small corpuscles or vacuoles are sometimes noted. The nucleolus stains with pyronine and other stains and absorbs ultraviolet light at 260 nm. Treatment with ribonuclease shows that the capacity to absorb this basophilic stain and ultraviolet ratiation depends on the presence of RNA.

The nucleolus may be surrounded by a ring of Feulgen-positive chromatin, which represents the heterochromatic regions of the chromosomes associated with the nucleolus. In large nucleoli some Feulgen-positive granules can be seen in portions of the chromosomes that penetrate the nucleolus. The nucleolus is formed at the nucleolar organizer, which is a chromosomal site that contains tandem repeats of the genes coding for 18S and 28S and rRNA. This DNA becomes uncoiled and penetrates the nucleolus, where it is actively transcribed. In addition, other ribosomal components, such as 5S RNA and the ribosomal proteins, which are synthesized in other parts of the cell, converge on the nucleolus, where the assembly of ribosomal subunits starts. Electron microscopy has confirmed the existence of a definite submicroscopic organization within the nucleolus. In some cells nucleoli have a compact structure; in others the structure may be more or less open and include clear regions that communicate with the nucleoplasm. Two characteristic components may be recognized: the granular zone and the fibrillar zone.

Granular Zone

The *granular zone* consists of electron-dense granules of 15 to 20 nm, *i.e.*, smaller in diameter than the ribosomes. Frequently the granular zone occupies the more peripheral region of the nucleolus, which is surrounded by the associated chromatin. Both the granular and the fibrillar zones are most clearly distinguished. The *fibrillar zone* consists of fine fibres 5 to 10 nm in diameter. It generally occupies the central region of the nucleolus. Both the granular and fibrillar zones are digested by ribonuclease. In some cells the nucleolus is surrounded by a ring of heterochromatin, called the *nucleolarassociated* chromatin, which sometimes extends into the nucleolus. Autoradiographic studies indicate that ^{3}H-uridine is incorporated first into the fibrillar area and is only later found in the granular part. This suggests the relationship:

$$\text{nucleolar DNA} \rightarrow \text{fibrillar area} \rightarrow \text{granular area}$$

The fibrillar area contains the ribosomal DNA and its initial transcripts (called 45S RNA), and the granular part contains ribosomal precursor particles at different stages of maturation. During mitosis nucleoli undergo cyclic changes. Nucleoli are formed around the DNA loop that extends from the nucleolar organizer. There may be several nucleoli per cell, but frequently they tend to fuse into one or few nucleoli. During late prophase the DNA loop containing the rRNA genes gradually retracts and coils into the nucleolar organizer of the corresponding chromosome. Since this DNA is very extended as a consequence of intense RNA synthesis, the nucleolar organizer region is one of the last to undergo condensation, thus producing a *secondary constriction* on the chromosome. The fibrillar and granular components are gradually dispersed into the nucleoplasm. After the cell divides, during telophase, the nucleolar organizer DNA uncoils and the nucleolus is reassembled.

A knowledge of the behaviour of the chromosomes during meiosis and especially in what has already been said about the genetic consequences. However, in 1856 when Gregor johann Mendel discoverd the fudamental laws of inheritance, nothing was known about the chromosomes and meiosis. His discovery was based on the precise quantification of experimental crosses and exceptional abstract thinking. He studied crosses between peas (*Pisum sativum*) which had pairs of differential or contrasing characteristics. For example, he used plant that have white and red flowers, smooth and roguh seeds, yellow and green seeds, long and short stems, and so forth. After crossing the parental generation (P_1), he observed the resulting *hybrids* of the first filial generation, F1. Then he crossed the hybrids (F_1) among themselves an studied the result in the second filial generation. F_2. In a cross between parents with yellow and green seeds in the first generation Mendel found that all the hybrids had yellow seeds and thus, the characteristic of only one parent In the second cross (F_2), the characteristics of both parents reappeared in the proportion of 75 percent to 25 per cent, or 3:1. Mendel postulated that the colour of the seeds was controlled by a "factor" that was transmitted to the offspring by means of the gametes. This hereditary factor, which is now called the *gene,* could be transmitted without mixing with othr genes. He postulated that the gene could be segregated in the hybrid into different games to be *distributed* in the offspring of the hybrid.

For this reason this is called the *law* or *principle of segregation of the genes.* Mendel found that the plants with yellow colour, had different genetic constutions. One-third of this group always gave yellow seeds, but the other two-thirds o the F_2 generation produced planats with yellow and green seeds in the ratio of 3:1 When the 25 per cent of plants in F_2 with green seeds wrer a pure strain for this character. If we represent the genes in the crossing by letters, designating by *A* the gene with yellow character and by *a* the gene with green character, we have the folloing.

In the first generation (F_1) both *A* and *a* genes are present, but only A is revealed because it is *dominant;* gene *a* remains hidden and is called *recessive.* In the hybrid F_1 both genes are segregated and enter different gametes. Half of them will have the gene *A,* and the other half,*a.* Since each individual produces two types of games in each sex, there are four possible combinations in $F_{2.}$ This gives as a result the proportion 1:2:1, corresponding to 25 per cent of plants with pure yellow seeds (*AA*), 50 per cent with hybrid yellow seed (*Aa*), and 25 per cent with pure green seeds (*aa*).

Mendel's results can now be explained in terms of the behavioural of chromosomes and genes. The genes present in the chromosomes are found in pairs called *alleles.* In each homologous chromosome the gene for each trait occurs at a particular point called a *locus* (plural *loci*). The gray mouse will have two GG genes, one in each homologue. Since the two homologues pair and then separate to enter the gametes. The mechanism is the same in a dominant as in a recessive white mouse. In the hybrid F_1, one chromosome bears gene G and the homologous chromosome bears gene *g.* When hybrids are self-fertilized, the gametes unite in the combinations shown by the checkerboard method illustrated in the individuals having two similar alleles are called *homozygous* (i. e., *GG* or *gg*); those with different alleles, *heterozygous.*

Genotype and Phenotype

Johanssen proposed the term *genotype* for the genetic constitution and *phenotype* for the visible characteristics shown by the individual in (91). For example, in the case of the peas with green or yellow seeds there are two phenotypes in $F_{2:}$ yellow seeds and green seeds in the ratio of 3:1, respectively. However, according to the genetic constitution, there are three different genotypes: *AA, Aa,* and *aa* in the ratio of 1:2:1.

The phenotype includes all characteristics of the individual that are an expression of gene activity. For example, in humans phenotypic characteristics include eye colour, baldness, the different hemoglobins, the blood groups, and a difference in the ability to taste thiourea. In crossings of certain plants that have white and red flowers, such as *Mirabilis jalapa,* it is possible to find in F_2 three phenotypes (red, pink, and white flowers), which correspond to the three genotypes. This is due to incomplete dominance. The principle of dominance and recessiveness in not always accomplished completely; dominance may be complete in most cases, but incomplete in others. In this case there is a mixture of characteristics, called *intermediary heredity.*

Whereas the law of segregation applies to the behaviour of a single pair of genes, the *principle of independent assortment* describes the simultaneous behaviour of two or more pairs of genes located in different pairs of chromosomes. Genes that lie in separate chromosomes are independently distributed during meiosis. The resulting off-spring is a hybrid (also called a dihybrid) at two loci.

All the above-mentioned examples of genetic crosses illustrated the fact that during meiosis there is a random distribution of the chromosomes which leads to the segregation of the genes in the gametes. However, when this type of study was carried out in the fruit fly *Drosophila melanogaster* by Morgan and collaborators, it became evident that the law of independent assortment was not universally applicable and that in certain crosses of two or more allelic pairs of genes, there was a certain limitation of free segregation. In each case there was a marked tendency for parental combinations to remain linked and for a lesser proportion of new combinations to be produced. In *Drosophila* there are only four pairs of chromosomes, and this increases the chances for genes to occupy loci in the same chromosome. If two genes A and B, with the corresponding alleles a and b, are in the same chromosome, only two classes of gametes will be obtained. After studying a considerable number of different crosses in *Drosophila,* Morgan reached the conclusion that all genes of this fly were clustered into four linked groups corresponding to the four pairs of chromosomes.

Recombination During Prophase

Further studies along these lines revealed that the linkage is not absolute and that it may be broken with a certain frequency. We have seen in the previous chapter that *genetic crossing over* or *recombination* takes place between the DNA molecules and is not visible microscopically. However, the *chiamata* observed at diplonema morphologically represent in some way the sites of these molecular events. Thus, there should be a correlation between the two phenomena. It is also known that the frequency of recombination of two linked genes is a function of the distance which separates them along the chromosome. When two genes are close to one another, the probability of crossing over is less than when they are far apart. If the distance between genes is estimated by linkage analysis, it is possible to construct a map indicating the relative position of each gene along the chromosome. An appraisal of the possible number of new recombinations may be obtained by counting the number of chiasmata during meiosis. The so-called *recombination index* is calculated by adding to the number of bivalents the number of chiasmata detected in the same cell at diplonema. In a species with a higher indes, the possibility of new recombinations is higher, and this implies a greater possibility of variation.

There are several other pieces of evidence to support the relationship between chiasmata and crossing over. The number of chiasmata is related to the length of the chromosomes in the bivalent. The presence of one chiasma reduces the possibility of another occurring in the immediate vicinity. This phenomenon has been called *positive interference.* The chiasma frequency is roughly constant in a given species, but can be modified by genetic or environmental action. Chromosome pairing, chiasma formation, and crossing over are under genetic control. Lines of rye with high and low chiasma frequencies have been obtained by inbreeding. Among environmental factors that may affect the number of chiasmata are temperature, radiation, chemicals, and nutrition.

Neurospora

Among the different organisms studies in genetics, the mold *Neurospora* osccupies a special place.

The advantage of this material is twofold:

- It is possible to identify and to follow the fate of each of the four chromatids present in the bivalent meiotic chromosome and thus to determine whether the crossing over involves two, three, or all four chromatids.
- It is possible to make a close correlation between genetic constitution and biochemical expression of genes.

The four cells resulting from the two meiotic division undergo a mitotic division, which gives rise to eight haploid ascospores. Each of these ascospore can be isolated by dissection and cultured separately, giving rise to haploid individuals having the genetic constitution carried in each of the four original chromatids of the bivalent chromosome.

Cytogenetics

The normal functioning of the genetic system is maintained by the constancy of the hereditary material carried in the chromosomes. We mentioned on Chapter all have already studied that the species has a diploid number of chromosomes which are in homologous pair, and the normal human karyotype will be studied in detail. Under various conditions there may be changes in the karyotype, with different genetic consequences. Chromosomes may *change in number,* maintaining a normal structure, or they may have *sructural alterations,* frequently called *chromosomal aberrations.* Such structural changes may appear spontaneously, or they may be caused by the action of ionizing rediations or by chemcals. Sometimes these changes may have a genetic component. For example, in the hereditary disease *xeroderma pigmentosum,* chromosomes are extremely sensitive to ultraviolet rediation.

Euploidy

There are two main kinds of change in the number of chromosomes. In *euploids,* the change is in the haploid number, but the entire set of chromosomes is kept balanced, whereas in *aneuploids* there is a loss or gain of one or more chromosomes causing the set to become unbalanced. In humans aneuploidy may cause severe alterations of the phenotype.

Some exceptional plants and animals have a *monoploid* (or *haploid)* chromosomes set. In these organisms meiosis is irregular because of the absence of homologous chromosomes. As a result, gametes with varying numbers of chromosomes may be formed.

Polyloidy

A plant or animal that has more than two haploid sets of chromosomes is called a *polyplid.* This change is common in nature especially in the flowering plants. A diploid organism has two similar genomes; a triploid has three; an autoterapoid has for; and so on. Polyploids may originate either by reduplication of the chromosome number inn somatic tissue with suppression of cytokinesis or by formation of gametes with an unreduced number of chromosomes. Meiosis in a triploid is more irrgular than in a tetraploid. In general, polyploids of uneven number are sterile because the gametes have a more unbalanced number of chromosomes.

The scarcity of polyploids among animals is due the mechanism by which sex is determined. If polyploidy occurs, the genic balance between the sexchromosomes and the autosomes is disturbed, and the race or species may disappear because of sterility.

Several species of amphibians show spontaneous polyploidy. In such polyploids the AND content is correspondingly increased. Polyploids are useful in the study of the expression of genes in multiple dosage. Studies of this type have been made on amphibians for serum albumin, hemoglobin, and various enzymes. Polyploidy has been induced experimentally by temperature shock. it is also possible to induce polyploidy with substances such as colchicine. acenaphthene, heteroauxin, and veratrine.

These substances inhibit the formation of the spindle, and thus cell division is not completed. After a time the cells recover their normal activity, but have double the number of chromosomes. Form the standpoint of pure and applied scientific work innumerable possibilities are offered by the experimental production of polypoids.

Allopolyploidy

This is a type of chromosomal variation that is produced in crosses between two species having different sets of chromosomes. The resulting hybrid has a different number of chromosomes. The resulting hybrid has a deferents. For example, the Argentine black *Sorghum (S. almum)* is an allotetraploid (2n=4×=40) originated in nature by an interspecific cross between *S. sudanense* (2n=2×=20). In this case the fertilization occurred between one abnormal diploid gamete of *S sudanense* and a normal gamete of the other species. In most cases crosses between distantly related species produce sterile diploid bybrids. *Aneuploidy* is produced by a failure in the separation of chromosomes during meiosis, called *non-disjunction. Monosomic* individuals have lost one of the chromosomes of the karyotype, and in *trisomics* there is a gain of one chromosome. These changes are studied specifically in the following chapter on the human karyotype.

Aneuploidy may also occure in somatic cells inn which the non-disjunction occurs at mitosis various tissues or parts of the body that have different chromosome numbers, *e.g.*, in mosaic individuals, variegation, and gynandromorphs.

In cultures of normal cells chromosomes variations are common, particularly after several transplants. These changes may lead to malignancy, but this is not the case in all cultures. The cytogenetic analysis of mammalian tumours has led investigators to consider them as altered karyotypes, which are genetically and cytologically unstable.

To understand the mechanism by which chromosomal aberrations are produced the student should recall the folded fibre model of chromosomes structure and the concept of *mononemic* structure of chromatids. According to this concept, each chromatid is made of one molecule of DNA; thus in the G_1 period there is a single DNA duplication that each chromosome contains two chromatids and two DNA molecules. In a chromatids and each chromosomal aberration there is an alteration in the structural organization of the chromosome.

For example, a break in the unit fibre may lead to a loss of material, called a *deletion,* and in cases inn which there are two breaks, a *translocation,* an *inversion,* or an *exchanges between sister chromatids* may be produced. In all these cases the chromosomes change can be observed microscopically, especially by observing the chromosomes at the stages of condensation. Chromosomal aberrations should be differentiated from *gene mutations* (also called *point mutations*), in which the change are produced at the molecular level. In these, the alteration resides in the genetic code (*i.e.*, in the sequence of DNA bases) and can be detected only by its genetic or molecular expression.

DNA

Both gene muations and chromosomal aberrations may occur spontaneously, but their frequency increases by the action of so-called ionizing rediation (*i.e.*, x-rays, B-reys, y-reys and ultraviolet light, as well as accelerated particles such as fast neutrons, protons, and soforth). (Other *mutagenic agents* are chemical substances, viruses, mycoplasms, and temperature changes. In various organisms it has been demonstrated that the number of mutations induced by radiation is proportional to the dose. In the case of *Drosophila,* the relation is a linear function in the range 25 to 9000 roentgen units (R).

The effects of radiation are cumulative over long periods of time. For example, 0.1 R per day for 10 years is enough to increase the mutation rate to about 150 per cent of the spontaneous level. In contrast to gene mutations, chromosomal aberrations increase exponentially. More chromosomal aberrations are produced by continuous than by intermittent treatment. A low dose of radiation may not be enough to cause fractures in one chromosome. As the dose increases, the number of breaks increases, and "aberrant" fusions becomes more and more likely, If the dose is intermittent or of low intermittent or of low intensity, there is a greater chance that the broken ends of the chromosome will rejoin, or "heal," in the original chromosome structure before a second break could cause aberration.

The type of chromosomal aberration depends on the period of the cell cycle the cell was in at the time of irradiation, A *chromatid break* occurs in the G_2 period, and *chromosomal break* (*i.e.*, two chromatids) occurs if the cell was in the G_1 period, which is before DNA duplication. Irradiation, generally produces localized lesions which may stabilize and from the so-called gaps. After some time, such gaps may be repaired. Such repair may result in complete restitution of the original structure (true repair). If there is no repair, the lesion becomes stabilized and cytologically visible. The process of restitution is inhibited by cold, cyanide, and dinitrophenol. Oxygen increases the number of fractures and chromosomal interchanges. During breakage of chromosome, the two fragments may either reunite or remain separated permanently. If, instead of reconstituting at once, the broken ends reduplicate to from two chromatids, the fragments with the centromere migrate to different poles, and the fragments without a centromere (acentric) are eliminated in the cytoplasm. Sometimes tho sister fragements reduplicate and unite, forming a dicentric chromosome and another acentric chromosome. During mitosis, the centromeres of the dicentric chromosome move to opposite poles and form a "bridge" between the daughter nuclei that finally break at some point.

In man, double of mutiple chromosomal breaks are induced by acute exposures (*e.g.*, heavy medical irradiations, nuclear accidents, or nuclear warfare), whereas single breaks are produced at low doses. Chromosomal aberrations have been observed inn blood cultures of humans who have had rediation treatments or injections of radioactive substances. The genetic effect of radiation has been studied in space flights. After Geminals III and IV. no increase in chromsomal aberrations was found in the blood cells of the astronauts who made the flights

Mutations in *somatic cells* are not transmitted from generation to generation but may be cumulative and produce severe changes in the individual, depending on the type of cell affected and the time at which the mutation occurs. Radiation can affect tissues that undergo mitosis, as well as tissues in which cell division no longer takes place. It mutation occurs during early embryonic development, a large number of cells are affected. The majority of mutated genes are recessive and thus no effect as long as the individual is heterozygous.

It is probable that some cases of cancer produced in irradiated individuals are caused by somatic mutation. In contrast to somatic mutation, mutations inn the germ cells may be transmitted to the offspring. However, in most cases of mutation caused by irradiation, both somatic and germ cells are frequently affected. Even the lowest doses of rediation are genetically harmful, and the effect are dangerous to all organisms, from the simplest to man. In microorganisma and plants, however, a few useful mutations may be obtained by irradiation. For example, irradiation is one method of producing new antibiotics and plants with high economic value.

Deficiency

Deletion or loss of chromosomal material may be either *terminal or interstitial,* originating from one or two breaks, respectively, produced in the G_1 period. Some interstitial deletions may originate both a ring chromosome and an acentric rod, which is eliminated.

In heterozygous deficiency, one chromosome is normal, but its homologue is deficient. Animals with a homozygous deficiency usually do not survive to an adult stage because a complete set of genes is lacking. This suggests that most genes are indispensable at least in a single dose to the development of a viable organism Deficiencies are important in cytogenetic investigations of gene location of for determination of the presence and position of unmatched genes.

Duplication

Duplication occurs when a segment of the chromosome is represented two or more times. The duplicated fragment may be free, with a centromere of its own, or it may be incorporated into a chromosomal segment of the normal complement, If the fragment includes the centromere, it may be incorporated as a small extra chromosome. In general, duplication are less deleterious to the individual than deficiencies.

Translocation

In the so-called *reciprocal translocation,* segment are exchanged between non-homologous chromosomes. Such translocation may be homozygotic or heterozygotic.

Cytologically, a translocated homozygote cannot be distinguished from a normal pair of chromosome, but it can be detected by genetic experiments. Heterozygotic translocations give rise to special pairing configurations in meiosis. An interesting result occurs when during translocation, both chromosome are broken vary close to their centromeres. The fusion creates a metanentric chromosome with two arms in the form of a V. and a small fragment, which tends to be eliminated.

Centric fusion has occurred during the phylogeny of *Drosophila,* grasshoppers, reptiles, birds, mammals, and other groups. It is a process that establishes a new type of chromosome and reduces the somatic chromosome number of the species. In the following chapter we will mention the importance of reciprocal translocations in human cytogenetics.

Inversion

An inversion is a chromosomal aberration in which a segment is inverted 180 degrees. Inversions are called pericenteric when the segment includes the centromere, and paracentric if the centromere is located outside the segment. In these aberrations there is typical configuration at pachynema consisting of a loop that allows the pairing of the inverted segement. In the paracentric inversion acentric and dicentric chromatids are formed.

Isochromosomes

A new type of chromosome may arise from a break (i. e., misdivision) at the centromere. The two resultant telocentric chromosomes may open up to produce chromosomes with two identical arms (i. e., isochromosome). This type of chromosome has been produced in irradiated material. At meiosis they may pair with themselves or with a normal homologue.

A *sister chromatid exchange* is an interchange of DNA between sister chromatids in chromosome, presumably involving DNA breakage followed by fusion. Sister chromatid exchanges are difficult to find using common cytologic methods because the chromatids are morphologically identical. Such chromatid exchanges were first described in studies in which ^{3}H-thymidine was added during a replicating cycle which was followed by another cycle in a non-radioactive medium. As was mentioned in earlier this method demonstrated that the duplication of DNA is semiconservative however, in a few mitoses observed by radioautography, an alteration of the labelling along the chromatids was found and this was interpreted as the result of chromatid exchange caused by the radioactivity of the tritiated label used. Analysis of this phenomenon has been greatly facilitated by the of *bromodeoxyuridine* (Brd U), a thymidine analogue that can be incorporated into the DNA of replicating cells instead of the original base. If BrdU is followed by a fluorescent day, the fluorescence of the segments that contain BrdU is greatly diminished in comparison with those with the original base. Furthermore, there is also similarly decreased staining with the Giemsa stain. The use of this technique, however, has been unable to discover whether the chromatid exchange could occur spontaneously or whether it is induced by the BrdU. It has, though, been of great help in differentiating the various inherited diseases characterized by chromosomal fragility, which have an increased frequency of sister chromatid exchanges and a tendency to have associated neoplasia. Some of these diseases (*e.g.*, Bloom's syndrome, Fanconis anemia, and ataxia-telangiectasia) are presumably related to defects in DNA repair.

Sister chromatid exchange has also been important inn studying the effect of mutagens on the chromosomes. Various mutagenic drugs that are alkylating agents, such as *mitomycin* C and nitrogen mustered, produce a great number of breaks and chromatid exchages. The intimate association of sister chromatid exchange with mutagenesis and carcinogenesis may have important medical implications.

Fundamental Role of Chromosome in Evolution

The development of comparative cytology and cytogenetics has brought about great progress towards an understanding of evolution. MuClung and S. Navashin were the first to emphasize the importance of cytogenetics to toxnomy and to the study of evolution by comparing genomes of related species. Systematics has been greatly advanced by cytogenetic investigation, which now provides many of the best methods for elucidating correlations between different taxonomic categories. In general, families, genera, and species are characterized by different genetic systems.

The study of the karyotypes of different species has revealed interesting facts about both the plant and animal kingdoms. It has been demonstrated that individuals in wild populations are, to some extent, heterogeneous cytologically and genetically. In some cases, even if the genes are identical they may be ordered in a different way, owing to alteration of the chromosomeal segments. These changes have ann important bearing on the evolution of species.

The majority of plant species originate from an abrupt and rapid change in nature, and aneuploidy and polyploidy are the prime sources of variation. In the animal kingdom polyploidy is not so important. Among vertebrates, different species of fishes have different number of chromosomes. Amphibians are generally characterized by a specific number for each family. Reptiles and bird have large chromosomes (macrochromosomes) and small chromosomes (microchr-omosomes) that serve to differentiate them cytologically.

Matthey distinguishes between the basic chromosome number and the number of chromosomal arms, also called the fundamental number (FN). According to this concept, the metacentric chromosome has *two* arms and acrocentric and metacentric chromosomes, and the number of arms in each the different species can be compared.

Another method used to study the cytogenetics of evolution is the application of measurements of total chromosomal area and DNA content.

With regard to the absolute size of their chromosomes, mammals and birds constitute two independent groups. These two orders have different DNA contents and different sexdetermining mechanisms. Speciation depends more of chromosomal rearrangement and mutation of individual genes than on changes in the total amount of genetic content.

Two opposite changes in the number (and configuration) of chromosomes are of particular importance in evolution. In *centric fusoin,* aprocess that leads to decrease in chromosomes join together to produce a metacentric chomosome. In *dissociation, or fission,* a process that leads to an increase in chromosome number, metacentric (commonly large) and a small supernumerary metacentric fragment become translocated, so that two acrocentric or submetacentric chromosomes are produced.

Fusion and dissociation are the main mechanisms by which the chromosome number can be decreased and increased during evolution of the majority of animals and in some groups of plants.

In recent years the karyotype of the great apes, chimpanzee, gorilla, and oragutan, have been compared with that of man. It was found that these primates have 48 chromosomes. and attempts were made to correlate the chromosome fairs with those of the 46 human chromosomes. This analysis has been reatly facilitated by the use of banding techniques. Using the quinacrine technique it was demonstrated that human chromosomes 1, 3, 5, 6, 7, 8, 10, 11, 12, 13, 14, 15, 19, 20, 21, 22, X and Y had direct counterparts in the chimpansee, but has less similarity with the paris of the gorills and orangutan. The difference between the 48 pair karyotype and the 46 is due to the fusion of two acrocentric chromosomes, present in the hominoid apes, to make the human chromosome 2[7].

At present, pericentric inversion appears to be the main structural difference between individual chromosomes of the great apes and man. It has been postulated that evolution of the lruman karyotype has occurred by series of percentric inversions which permitted genetic isolation of small breeding groups and selection of favourable gene combinations that gave rest to *Home sapiens*. The homologies in morphology of the chromosomes are now being correlated with the mapping of genes in similar chromosome of man and the great apes. Studies of somatic and polytene chromosomes in several hundred species of *Drosophila* have elucidated the formulation and evolution of this elution of this genus, which has been thoroughly analysed from, genetic, ecologic, and geographic standpoints. Observation of chromosomal organization and of the different karyotypes in the individual, the species genera, and the major systematic groups indicates that a chromosomal mechanism is involved in the process of evolution.

The problem of evolution, however, is very complex and should be considered from the different biochemical, cytologic, genetic, ecologic, and experimental aspects. All these methods and approaches should be used to analyse the intricate relationships between groups off organisms, particularly those that show marked variations.

Functions of Chromosomes

The discovery that DNA is the prime genetic molecule, carrying all the hereditary information within chromosomes, immediately focused attention on its structure. It was hoped that knowledge of the structure would reveal how DNA carries the genetic messages that are replicated when chromosomes divide to produce two identical copies of themselves. During the late 1940s and early 1950s, several research groups in the United States and in Europe engaged in serious efforts—both cooperative and rival—to understand how the atoms of DNA are linked together by covalent bonds and how the resulting molecules are arranged in three-dimensional space. Not surprisingly, there initially were fears that DNA might have very complicated and perhaps bizarre structures that differed radically from one gene to another.

Great relief, if not general elation, was thus expressed when the fundamental DNA structure was found to be the double helix. It told us that all genes have roughly the same three-dimensional form and that the differences between two genes reside in the order and number of their four nucleotide building blocks along the complementary strands.

Now, some 50 years after the discovery of the double helix, this simple description of the genetic material remains true and has not had to be appreciably altered to accommodate new findings. Nevertheless, we have come to realise that the structure of DNA is not quite as uniform as was first thought. For example, the chromosome of some small viruses have single-stranded, not double-stranded, molecules. Moreover, the precise orientation of the base pairs varies slightly from base pair to base pair in a manner that is influenced by the local DNA sequence.

Some DNA sequences even permit the double helix to twist in the left-handed sense, as opposed to the right-handed sense originally formulated for DNA's general structure. And while some DNA molecules are linear, others are circular.

Still additional complexity comes from the supercoiling (further twisting) of the double helix, often around cores of DNA-binding proteins. Likewise, we now realise that RNA, which at first glance appears to be very similar to DNA, has its own distinctive structural features. It is principally found as a single-stranded molecule.

Yet by means of intra-strand base pairing, RNA exhibits extensive double-helical character and is capable of folding into a wealth of diverse tertiary structures. These structures are full of surprises, such as non-classical base pairs, base-backbone interactions, and knot-like configurations.

Most remarkable of all, and of profound evolutionary significance, some RNA molecules are enzymes that carry out reactions that are at the core of information transfer from nucleic acid to protein. Clearly, the structures of DNA and RNA are richer and more intricate than was at first appreciated. Indeed, there is no one generic structure for DNA and RNA.

Polynucleotide Chains

The most important feature of DNA is that it is usually composed of two polynucleotide chains twisted around each other in the form of a double helix. The upper part presents the structure of the double helix shown in a schematic form. Note that if inverted 180°, the double helix looks superficially the same, due to the complementary nature of the two DNA strands.

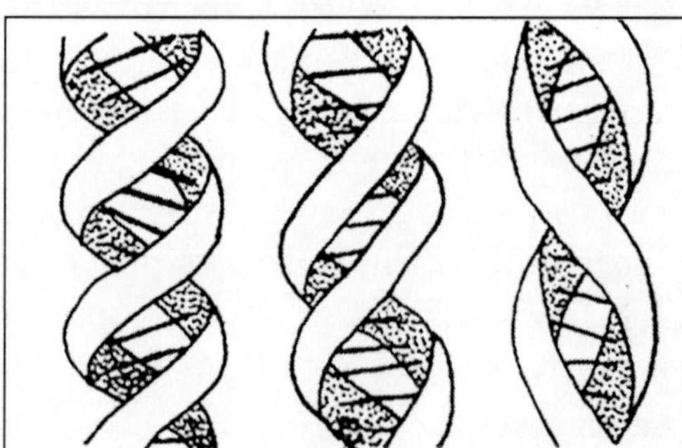

From	A	B	Z
Belical Twist	Right	Right	Left
Base Pairs Per Turn	11	10	12
Occurrence	RNA, DNA	DNA	DNA

Figure: *Polynucleotide Chains*

The space-filling model of the double helix, shows the components of the DNA molecule and their relative positions in the helical structure. The backbone of each strand of the helix is composed of alternating sugar and phosphate residues; the bases project inward but are accessible through the major and minor grooves.

Let us begin by considering the nature of the nucleotide, the fundamental building block of DNA. The nucleotide consists of a phosphate joined to a sugar, known as 2'-deoxyribose, to which a base is attached. The sugar is called *2'*-deoxyribose because there is no hydroxyl at position 2' (just two hydrogens).

Note that the positions on the ribose are designated with primes to distinguish them from positions on the bases. We can think of how the base is joined to 2'-deoxyribose by imagining the removal of a molecule of water between the hydroxyl on the 1' carbon of the sugar and the base to form a glycosidic bond. The sugar and base alone are called a nucleoside. Likewise, we can imagine linking the phosphate to 2'-deoxyribose by removing a water molecule from between the phosphate and the hydroxyl on the 5' carbon to make a 5' phosphomonoester. Adding a phosphate (or more than one phosphate) to a nucleo*s*ide creates a nucleo*t*ide. Thus, by making a glycosidic bond between the base and the sugar, and by making a phosphoester bond between the sugar and the phosphoric acid, we have created a nucleotide.

Nucleotides are, in turn, joined to each other in polynucleotide chains through the 3' hydroxyl of 2'-deoxyribose of one nucleotide and the phosphate attached to the 5' hydroxyl of another nucleotide. This is a phosphodiester linkage in which the phosphoryl group between the two nucleotides has one sugar esterified to it through a 3' hydroxyl and a second sugar esterified to it through a 5' hydroxyl. Phosphodiester linkages create the repeating, sugar-phosphate backbone of the polynucleotide chain, which is a regular feature of DNA. In contrast, the order of the bases along the polynucleotide chain is irregular. This irregularity as well as the long length is, as we shall see, the basis for the enormous information content of DNA. The phosphodiester linkages impart an inherent polarity to the DNA chain.

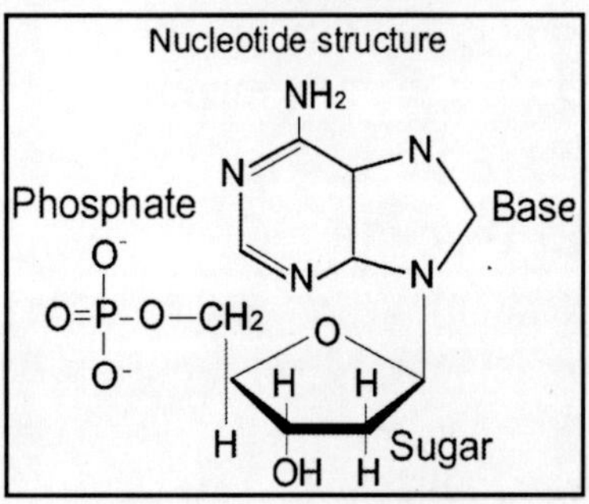

Figure: *Nucleotide*

This polarity is defined by the asymmetry of the nucleotides and the way they are joined. DNA chains have a free 5' phosphate or 5' hydroxyl at one end and a free 3' phosphate or 3' hydroxyl at the other end. The convention is to write DNA sequences from the 5' end (on the left) to the 3' end, generally with a 5' phosphate and a 3' hydroxyl.

Purines and Pyrimidines

The bases in DNA fall into two classes, purines and pyrimidines. The purines are adenine and guanine, and the pyrimidines are cytosine and thymine. The purines are derived from the double-ringed structure. Adenine and guanine share this essential structure but with different groups attached. The bases are attached to the deoxyribose by glycosidic linkages at N1 of the pyrimidines or at N9 of the purines.

Each of the bases exists in two alternative tautomeric states, which are in equilibrium with each other. The equilibrium lies far to the side of the conventional structures, which are the predominant states and the ones important for base pairing. The nitrogen atoms attached to the purine and pyrimidine rings are in the amino form in the predominant state and only rarely assume the imino configuration. Likewise, the oxygen atoms attached to the guanine and thymine normally have the keto form and only rarely take on the enol configuration. As we shall see, the capacity to form an alternative tautomer is a frequent source of errors during DNA synthesis.

Double Helix

The double helix is composed of two polynucleotide chains that are held together by weak, non-covalent bonds between pairs of bases. Adenine on one chain is always paired with thymine on the other chain and, likewise, guanine is always paired with cytosine. The two strands have the same helical geometry but base pairing holds them together with the opposite polarity.

That is, the base at the 5' end of one strand is paired with the base at the 3' end of the other strand. The strands are said to have an anti-parallel orientation. This anti-parallel orientation is a stereochemical consequence of the way that adenine and thymine and guanine and cytosine pair with each together.

Pairing of Adenine and Thymine

The pairing between adenine and thymine and between guanine and cytosine results in a complementary relationship between the sequence of bases on the two intertwined chains and gives DNA its self-encoding character. For example, if we have the sequence 5'-ATGTC-3' on one chain, the opposite chain must have the complementary sequence 3'-TACAG-5'. The strictness of the rules for this "Watson-Crick" pairing derives from the complementarity both of shape and of hydrogen bonding properties between adenine and thymine and between guanine and cytosine.

Adenine and thymine match up so that a hydrogen bond can form between the exocyclic amino group at C6 on adenine and the carbonyl at C4 in thymine; and likewise, a hydrogen bond can form between N1 of adenine and N3 of thymine.

A corresponding arrangement can be drawn between a guanine and a cytosine, so that there is both hydrogen bonding and shape complementarity in this base pair as well.

A G:C base pair has three hydrogen bonds, because the exocyclic NH_2 at C_2 on guanine lies opposite to, and can hydrogen bond with, a carbonyl at C_2 on cytosine. Likewise, a hydrogen bond can form between N_1 of guanine and N_3 of cytosine and between the carbonyl at C_6 of guanine and the exocyclic NH_2 at C_4 of cytosine.

Watson-Crick base pairing requires that the bases are in their preferred tautomeric states. An important feature of the double helix is that the two base pairs have exactly the same geometry; having an A:T base pair or a G:C base pair between the two sugars does not perturb the arrangement of the sugars. Neither does T:A or C:G. In other words, there is an approximately twofold axis of symmetry that relates the two sugars and all four base pairs can be accommodated within the same arrangement without any distortion of the overall structure of the DNA.

Bonding of the Double Helix

The hydrogen bonds between complementary bases are a fundamental feature of the double helix, contributing to the thermodynamic stability of the helix and providing the information content and specificity of base pairing. Hydrogen bonding might not at first glance appear to contribute importantly to the stability of DNA for the following reason.

An organic molecule in aqueous solution has all of its hydrogen bonding properties satisfied by water molecules that come on and off very rapidly. As a result, for every hydrogen bond that is made when a base pair forms, a hydrogen bond with water is broken that was there before the base pair formed. Thus, the net energetic contribution of hydrogen bonds to the stability of the double helix would appear to be modest. However, when polynucleotide strands are separate, water molecules are lined up on the bases. When strands come together in the double helix, the water molecules are displaced from the bases. This creates disorder and increases entropy, thereby stabilizing the double helix. Hydrogen bonds are not the only force that stabilizes the double helix.

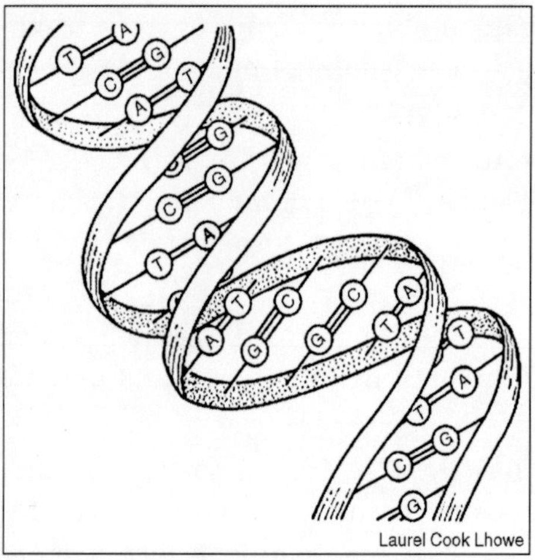

Figure: *Hydrogen Bonds Between Double Helix*

A second important contribution comes from stacking interactions between the bases. The bases are flat, relatively waterinsoluble molecules, and they tend to stack above each other roughly perpendicular to the direction of the helical axis. Electron cloud interactions (ð–ð) between bases in the helical stacks contribute significantly to the stability of the double helix. Hydrogen bonding is also important for the specificity of base pairing. Suppose we tried to pair an adenine with a cytosine.

Then we would have a hydrogen bond acceptor (N_1 of adenine) lying opposite a hydrogen bond acceptor (N_3 of cytosine) with no room to put a water molecule in between to satisfy the two acceptors. Likewise, two hydrogen bond donors, the NH_2 groups at C_6 of adenine and C_4 of cytosine, would lie opposite each other. Thus, an A:C base pair would be unstable because water would have to be stripped off the donor and acceptor groups without restoring the hydrogen bond formed within the base pair.

Left-Handed Helix

DNA containing alternative purine and pyrimidine residues can fold into left-handed as well as right-handed helices. To understand how DNA can form a left-handed helix, we need to consider the glycosidic bond that connects the base to the 1' position of 2'-deoxyribose. This bond can be in one of two conformations called *syn* and *anti*.

In right-handed DNA, the glycosidic bond is always in the *anti* conformation. In the left-handed helix, the fundamental repeating unit usually is a purine-pyrimidine dinucleotide, with the glycosidic bond in the *anti* conformation at pyrimidine residues and in the *syn* conformation at purine residues. It is this *syn* conformation at the purine nucleotides that is responsible for the left-handedness of the helix. The change to the *syn* position in the purine residues to alternating *anti–syn* conformations gives the backbone of left-handed DNA a zigzag look, which distinguishes it from right-handed forms. The rotation that effects the change from *anti* to *syn* also causes the ribose group to undergo a change in its pucker. In solution alternating purine–pyrimidine residues assume the left-handed conformation only in the presence of high concentrations of positively charged ions (*e.g.*, Na^+) that shield the negatively charged phosphate groups. At lower salt concentrations, they form typical right-handed conformations. The physiological significance of Z DNA is uncertain and left-handed helices probably account at most for only a small of proportion of a cell's DNA.

DNA Strands

Because the two strands of the double helix are held together by relatively weak (non-covalent) forces, we might expect that the two strands could come apart easily. Indeed, the original structure for the double helix suggested that DNA replication would occur in just this manner. The complementary strands of double helix can also be made to come apart when a solution of DNA is heated above physiological temperatures (to near 100°C) or under conditions of high pH, a process known as denaturation. However, this complete separation of DNA strands by denaturation is reversible. When heated solutions of denatured DNA are slowly cooled, single strands often meet their complementary strands and reform regular double helices. The capacity to renature denatured DNA molecules permits artificial hybrid DNA molecules to be formed by slowly cooling mixtures of denatured DNA from two different sources.

Likewise, hybrids can be formed between complementary strands of DNA and RNA. The ability to form hybrids between two single-stranded nucleic acids (hybridization) is the basis for several indispensable techniques in molecular biology, such as Southern blot hybridization and DNA microarrays.

Important insights into the properties of the double helix were obtained from classic experiments carried out in the 1950s in which the denaturation of DNA was studied under a variety of conditions. In these experiments DNA denaturation was monitored by measuring the absorbance of ultraviolet light passed through a solution of DNA.

DNA maximally absorbs ultraviolet light at a wavelength of about 260 nm. It is the bases that are principally responsible for this absorption. When the temperature of a solution of DNA is raised to near the boiling point of water, the optical density (absorbance) at 260 nm markedly increases. The explanation for this increase is that duplex DNA is hypochromic; it absorbs less ultraviolet light by about 40 per cent than do individual DNA chains. The hypochromicity is due to base stacking, which diminishes the capacity of the bases in duplex DNA to absorb ultraviolet light. If we plot the optical density of DNA as a function of temperature, we observe that the increase in absorption occurs abruptly over a relatively narrow temperature range. The midpoint of this transition is the melting point or Tm.

Like ice, DNA melts: it undergoes a transition from a highly ordered double-helical structure to a much less ordered structure of individual strands. The sharpness of the increase in absorbance at the melting temperature tells us that the denaturation and renaturation of complementary DNA strands is a highly cooperative, zippering-like process.

Renaturation, for example, probably occurs by means of a slow nucleation process in which a relatively small stretch of bases on one strand find and pair with their complement on the complementary strand. The remainder of the two strands then rapidly zipper-up from the nucleation site to reform an extended double helix.

The melting temperature of DNA is a characteristic of each DNA that is largely determined by the G:C content of the DNA and the ionic strength of the solution. The higher the per cent of G:C base pairs in the DNA (and hence the lower the content of A:T base pairs), the higher the melting point. Likewise, the higher the salt concentration of the solution the greater the temperature at which the DNA denatures.

How do we explain this behaviour? G:C base pairs contribute more to the stability of DNA than do A:T base pairs because of the greater number of hydrogen bonds for the former (three in a G:C base pair versus two for A:T) but also importantly because the stacking interactions of G:C base pairs with adjacent base pairs are more favorable than the corresponding interactions of A:T base pairs with their neighboring base pairs. The effect of ionic strength reflects another fundamental feature of the double helix. The backbones of the two DNA strands contain phosphoryl groups, which carry a negative charge. These negative charges are close enough across the two strands that if not shielded they tend to cause the strands to repel each other, facilitating their separation. At high ionic strength, the negative charges are shielded by cations, thereby stabilizing the helix. Conversely, at low ionic strength the unshielded negative charges render the helix less stable.

DNA Molecules

It was initially believed that all DNA molecules are linear and have two free ends. Indeed, the chromosomes of eukaryotic cells each contain a single (extremely long) DNA molecule. But now we know that some DNAs are circles. For example, the chromosome of the small monkey DNA virus SV40 is a circular, double-helical DNA molecule of about 5,000 base pairs.

Also, most (but not all) bacterial chromosomes are circular; *E. coli* has a circular chromosome of about 5 million base pairs. Additionally, many bacteria have small autonomously replicating genetic elements known as plasmids, which are generally circular DNA molecules.

Interestingly, some DNA molecules are sometimes linear and sometimes circular. The most well-known example is that of the bacteriophage ‘, a DNA virus of *E. coli*. The phage ‘genome is a linear double-stranded molecule in the virion particle. However, when the ‘genome is injected into an *E. coli* cell during infection, the DNA circularizes. This occurs by base-pairing between single-stranded regions that protrude from the ends of the DNA and that have complementary sequences (“sticky ends”).

DNA Topology

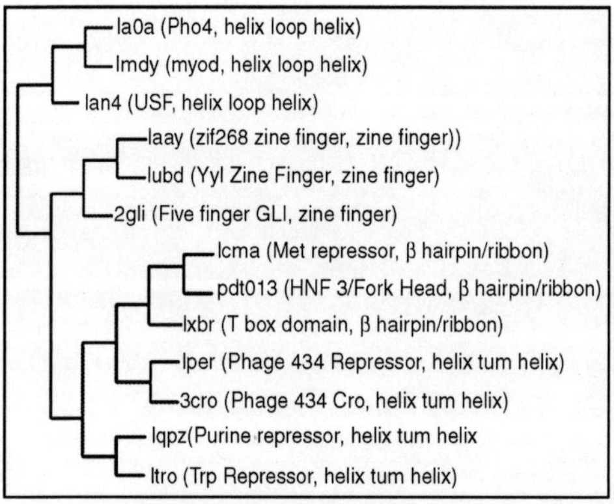

Figure: *DNA-Proteins*

As DNA is a flexible structure, its exact molecular parameters are a function of both the surrounding ionic environment and the nature of the DNA-binding proteins with which it is complexed. Because their ends are free, linear DNA molecules can freely rotate to accommodate changes in the number of times the two chains of the double helix twist about each other.

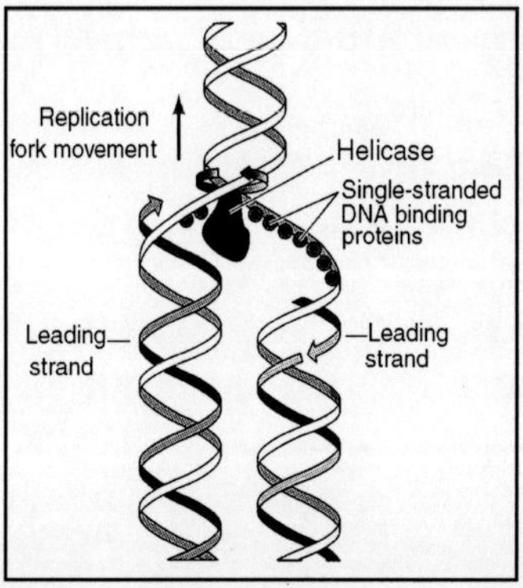

Figure: *DNA-binding Proteins*

But if the two ends are covalently linked to form a circular DNA molecule and if there are no interruptions in the sugar phosphate backbones of the two strands, then the absolute number of times the chains can twist about each other cannot change. Such a covalently closed, circular DNA is said to be topologically constrained. Even the linear DNA molecules of eukaryotic chromosomes are subject to topological constraints due to their entrainment in chromatin and interaction with other cellular components. Despite these constraints, DNA participates in numerous dynamic processes in the cell. For example, the two strands of the double helix, which are twisted around each other, must rapidly separate in order for DNA to be duplicated and to be transcribed into RNA.

Thus, understanding the topology of DNA and how the cell both accommodates and exploits topological constraints during DNA replication, transcription, and other chromosomal transactions is of fundamental importance in molecular biology.

Circular DNA

Let us consider the topological properties of covalently closed, circular DNA, which is referred to as cccDNA. Because there are no interruptions in either polynucleotide chain, the two strands of cccDNA cannot be separated from each other without the breaking of a covalent bond.

If we wished to separate the two circular strands without permanently breaking any bonds in the sugar phosphate backbones,we would have to pass one strand through the other strand repeatedly. The number of times one strand would have to be passed through the other strand in order for the two strands to be entirely separated from each other is called the linking number. The linking number, which is always an integer, is an invariant topological property of cccDNA, no matter how much the DNA molecule is distorted.

The linking number is the sum of two geometric components called the twist and the writhe. Let us consider twist first. Twist is simply the number of helical turns of one strand about the other, that is, the number of times one strand completely wraps around the other strand. Consider a cccDNA that is lying flat on a plane.

In this flat conformation, the linking number is fully composed of twist. Indeed, the twist can be easily determined by counting the number of times the two strands cross each other. The helical crossovers (twist) in a right-handed helix are defined as positive such that the linking number of DNA will have a positive value. But cccDNA is generally not lying flat on a plane.

Rather, it is usually torsionally stressed such that the long axis of the double helix crosses over itself, often repeatedly, in three-dimensional space. This is called *writhe*. To visualize the distortions caused by torsional stress, think of the coiling of a telephone cord that has been overtwisted. Writhe can take two forms. One form is the interwound or plectonemic writhe, in which the long axis is twisted around itself. The other form of writhe is a toroid or spiral in which the long axis is wound in a cylindrical manner, as often occurs when DNA wraps around protein. The writhing number (*Wr*) is the total number of interwound and/or spiral writhes in cccDNA. Interwound writhe and spiral writhe are topologically equivalent to each other and are readily interconvertible geometric properties of cccDNA. Also, twist and writhe are interconvertible. A molecule of cccDNA can readily undergo distortions that convert some of its twist to writhe or some of its writhe to twist without the breakage of any covalent bonds. The only constraint is that the sum of the twist number (*Tw*) and the writhing number (*Wr*) must remain equal to the linking number (*Lk*). This constraint is described by the equation: $Lk = Tw + Wr$.

Linking of Fully Relaxed cccDNA

Consider cccDNA that is free of supercoiling (that is, it is said to be relaxed) and whose twist corresponds to that of the B form of DNA in solution under physiological conditions (about 10.5 base pairs per turn of the helix). The linking number (*Lk*) of such cccDNA under physiological conditions is assigned the symbol Lk^o. Lk^o for such a molecule is the number of base pairs divided by 10.5.

For a cccDNA of 10,500 base pairs, Lk = +1,000. (The sign is positive because the twists of DNA are right-handed.) One way to see this is to imagine pulling one strand of the 10,500 base pair cccDNA out into a flat circle. If we did this, then the other strand would cross the flat circular strand 1,000 times. How can we remove supercoils from cccDNA if it is not already relaxed? One procedure is to treat the DNA mildly with the enzyme DNase I, so as to break on average one phosphodiester bond (or a small number of bonds) in each DNA molecule. Once the DNA has been "nicked" in this manner, it is no longer topologically constrained and the strands can rotate freely, allowing writhe to dissipate. If the nick is then repaired, the resulting cccDNA molecules will be relaxed and will have on average an *Lk* that is equal to Lk^o. (Due to rotational fluctuation at the time the nick is repaired, some of the resulting cccDNAs will have an *Lk* that is somewhat greater than Lk^o and others will have an *Lk* that is somewhat lower. Thus, the relaxation procedure will generate a narrow spectrum of topoisomers whose average *Lk* is equal to Lk^o).

Supercoiling in Eukaryotes

DNA in the nucleus of eukaryotic cells is packaged in small particles known as nucleosomes in which the double helix is wrapped almost two times around the outside circumference of a protein core. We will be able to recognize this wrapping as the toroid or spiral form of writhe. Importantly, it occurs in a lefthanded manner. It turns out that writhe in the form of left-handed spirals is equivalent to negative supercoils. Thus, the packaging of DNA into nucleosomes introduces negative superhelical density.

Topoisomerases

As we have seen, the linking number is an invariant property of DNA that is topologically constrained. It can only be changed by introducing interruptions into the sugar-phosphate backbone. A remarkable class of enzymes known as topoisomerases are able to do just that by introducing transient nicks or breaks into the DNA. Topoisomerases are of two broad types. Type II topoisomerases change the linking number in steps of two. They make transient double-stranded breaks in the DNA, through which they pass a region of uncut duplex DNA before resealing the break.

Type II topoisomerases require energy from ATP hydrolysis for their action. Type I topoisomerases, in contrast, change the linking number of DNA in steps of one. They make transient singlestranded breaks in the DNA, allowing one strand to pass through the break in the other before resealing the nick. Type I topoisomerases relax DNA by removing supercoils. They can be compared to the protocol of introducing nicks into cccDNA with DNase and then repairing the nicks, which as we saw can be used to relax cccDNA, except that type I topoisomerases relax DNA in a controlled and concerted manner. In contrast to type II topoisomerases, type I topoisomerases do not require ATP. Both type I and type II topoisomerases work through an intermediate in which the enzyme is covalently attached to one end of the broken DNA.

Both prokaryotes and eukaryotes have type I and type II topoisomerases, which are capable of removing supercoils from DNA. In addition, however, prokaryotes have a special type II topoisomerase known as DNA gyrase that is able to introduce negative supercoils, rather than remove them. DNA gyrase is responsible for the negative supercoiling of chromosomes in prokaryotes, which facilitates unwinding of the DNA duplex during transcription and DNA replication.

Electrophoresis

Covalently closed, circular DNA molecules of the same length but of different linking numbers are called DNA topoisomers. Even though topoisomers have the same molecular weight, they can be separated from each other by electrophoresis through a gel of agarose. The basis for this separation is that the greater the writhe the more compact the shape of a cccDNA. Once again, think of how supercoiling a telephone cord causes it to become more compact. The more compact the DNA, the more easily (up to a point) it is able to migrate through the gel matrix. Thus, a fully relaxed cccDNA migrates more slowly than a highly supercoiled topoisomer of the same circular DNA. Molecules in adjacent rungs of the ladder differ from each other by a linking number difference of just one. Obviously, electrophoretic mobility is highly sensitive to the topological state of DNA.

Helical Periodicity

The observation that DNA topoisomers can be separated from each other electrophoretically is the basis for a simple experiment that proves that DNA has a helical periodicity of about 10.5 base pairs per turn in solution. Consider three cccDNAs of sizes 3990, 3995, and 4011 base pairs that were relaxed to completion by treatment with topoisomerase I. When subjected to electrophoresis through agarose, the 3990- and 4011-base-pair DNAs exhibit essentially identical mobilities.

Due to thermal fluctuation, topoisomerase treatment actually generates a narrow spectrum of topoisomers, but for simplicity let us consider the mobility of only the most abundant topoisomer (that corresponding to the cccDNA in its most relaxed state). The mobilities of the most abundant topoisomers for the 3990- and 4011-base-pair DNAs are indistinguishable because the 21-base-pair difference between them is negligible compared to the sizes of the rings. The most abundant topoisomer for the 3995-base-pair ring, however, is found to migrate slightly more rapidly than the other two rings even though it is only 5 base pairs larger than the 3990-base-pair ring.

How are we to explain this anomaly? The 3990- and 4011- base-pair rings in their most relaxed states are expected to have linking numbers equal to Lk^{o}, that is, 380 in the case of the 3990-base-pair ring (dividing the size by 10.5 base pairs) and 382 in the case of the 4011-base-pair ring. Because Lk is equal to Lk^{o}, the linking difference ('$Lk = Lk - Lk^{o}$) in both cases is zero and there is no writhe. But because the linking number must be an integer, the most relaxed state for the 3995-base-pair ring would be either of two topoisomers having linking numbers of 380 or 381. However, Lk^{o} for the 3995-base-pair ring is 380.5.

Thus, even in its most relaxed state, a covalently closed circle of 3995 base pairs would necessarily have about half a unit of writhe (its linking difference would be 0.5), and hence it would migrate more rapidly than the 3990- and 4011-base-pair circles. In other words, to explain how rings that differ in length by 21 base pairs (two turns of the helix) have the same mobility whereas a ring that differs in length by only 5 base pairs (about half a helical turn) exhibits a different mobility, we must conclude that DNA in solution has a helical periodicity of about 10.5 base pairs per turn.

Ethidium Ions Cause DNA to Unwind

Ethidium is a large, flat, multi-ringed cation. Its planar shape enables ethidium to slip (intercalate) between the stacked base pairs of DNA. Because it fluoresces when exposed to ultraviolet light, and because its fluorescence increases dramatically after intercalation, ethidium is used as a stain to visualize DNA. When an ethidium ion intercalates between two base pairs, it causes the DNA to unwind by 26°, reducing the normal rotation per base pair from 36° to 10°. In other words, ethidium decreases the twist of DNA.

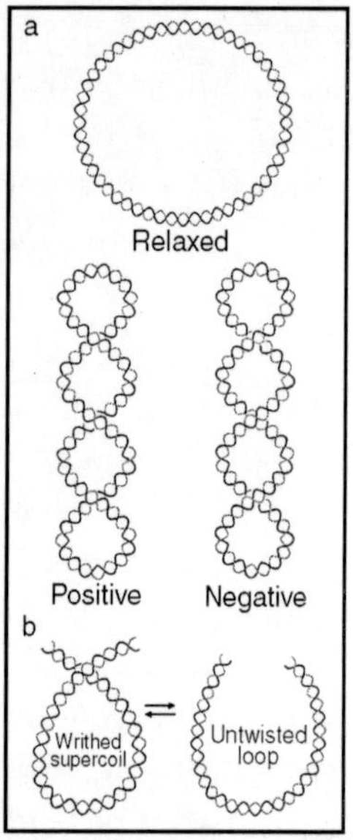

Figure: *Supercoiled DNA*

Imagine the extreme case of a DNA molecule that has an ethidium ion between every base pair. Instead of 10 base pairs per turn it would have 36! When ethidium binds to linear DNA or to a nicked circle, it simply causes the helical pitch to increase. But consider what happens when ethidium binds to covalently closed, circular DNA. The linking number of the cccDNA does not change (no covalent bonds are broken and resealed), but the twist decreases by 26° for each molecule of ethidium that has bound to the DNA.

Because $Lk = Tw + Wr$, this decrease in Tw must be compensated for by a corresponding increase in Wr. If the circular DNA is initially negatively supercoiled (as is normally the case for circular DNAs isolated from cells), then the addition of ethidium will increase Wr.

In other words, the addition of ethidium will relax the DNA. If enough ethidium is added, the negative supercoiling will be brought to zero, and if even more ethidium is added, Wr will increase above zero and the DNA will become positively supercoiled. Because the binding of ethidium increases Wr, its presence greatly affects the migration of cccDNA during gel electrophoresis. In the presence of non-saturating amounts of ethidium, negatively supercoiled circular DNAs are more relaxed and migrate more slowly, whereas relaxed cccDNAs become positively supercoiled and migrate more rapidly.

6

DNA and RNA

Deoxyribonucleic Acid

Deoxyribonucleic acid, or DNA, is a nucleic acid that contains the genetic instructions used in the development and functioning of all known living organisms (with the exception of RNA viruses). The main role of DNA molecules is the long-term storage of information. DNA is often compared to a set of blueprints, like a recipe or a code, since it contains the instructions needed to construct other components of cells, such as proteins and RNA molecules. The DNA segments that carry this genetic information are called genes, but other DNA sequences have structural purposes, or are involved in regulating the use of this genetic information.

DNA consists of two long polymers of simple units called nuclotides, with backbones made of sugars and phosphate groups joined by ester bonds. These two strands run in opposite directions to each other and are therefore anti-parallel. Attached to each sugar is one of four types of molecules called bases. It is the sequence of these four bases along the backbone that encodes information. This information is read using the genetic code, which specifies the sequence of the amino acids within proteins. The code is read by copying stretches of DNA into the related nucleic acid RNA, in a process called transcription.

Within cells, DNA is organized into long structures called chromosomes. These chromosomes are duplicated before cells divide, in a process called DNA replication. Eukaryotic organisms (animals, plants, fungi, and protists) store most of their DNA inside the cell nucleus and some of their DNA in organelles, such as mitochondria or chloroplasts. In contrast, prokaryotes (bacteria and archaea) store their DNA only in the cytoplasm. Within the chromosomes, chromatin proteins such as histones compact and organize DNA. These compact structures guide the interactions between DNA and other proteins, helping control which parts of the DNA are transcribed. Deoxyribonucleic acid is pretty unusual in that it is about the only common molecule capable of directing its own synthesis.

The processes of mitosis and meiosis were discovered in the 1870s and 1890s. It was observed that, as cells divided, chromosomes moved around in a cell, and people began to wonder what their function was. It was determined that chromosomes were made of protein and DNA, about which people knew almost nothing. People began to suspect that chromosomes had something to do with genetics, but couldn't explain what/how.

When enough evidence was accumulated to confirm that chromosomes did, indeed, have something to do with genetics, most people thought that in some way the protein in the chromosomes served as the genetic material. People knew that DNA was also in the chromosomes, but because its structure was unknown and people didn't know much about it, few people thought it was the genetic material. Frederick Griffith performed an experiment using pneumonia bacteria and mice. This was one of the first experiments that hinted that DNA was the genetic code material.

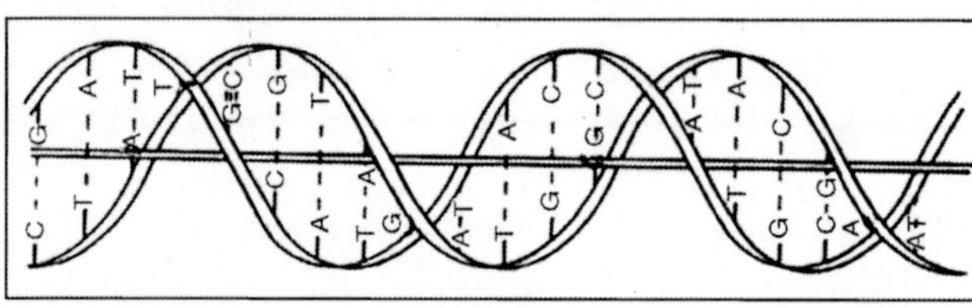

Figure: *DNA*

He used two strains of *Streptococcus pneumoniae*: a "smooth" strain which has a polysaccharide coating around it that makes it look smooth when viewed with a microscope, and a "rough" strain which doesn't have the coating, thus looks rough under the microscope. When he injected live S strain into mice, the mice contracted pneumonia and died. When he injected live R strain, a strain which typically does not cause illness, into mice, as predicted they did not get sick, but lived.

Thinking that perhaps the polysaccharide coating on the bacteria somehow caused the illness and knowing that polysaccharides are not affected by heat, Griffith then used heat to kill some of the S strain bacteria and injected those dead bacteria into mice. This failed to infect/kill the mice, indicating that the polysaccharide coating was not what caused the disease, but rather, something within the living cell. Since Griffith had used heat to kill the bacteria and heat denatures protein, he next hypothesized that perhaps some protein within the living cells, that was denatured by the heat, caused the disease. He then injected another group of mice with a mixture of heat-killed S and live R, and the mice died! When he did a necropsy on the dead mice, he isolated live S strain bacteria from the corpses. Griffith concluded that the live R strain bacteria must have absorbed genetic material from the dead S strain bacteria, and since heat denatures protein, the protein in the bacterial chromosomes was not the genetic material. This evidence pointed to DNA as being the genetic material. Transformation is the process whereby one strain of a bacterium absorbs genetic material from another strain of bacteria and "turns into" the type of bacterium whose genetic material it absorbed. Because DNA was so poorly understood, scientists remained skeptical up through the 1940s.

Alfred Hershey and Martha Chase did an experiment which is so significant, it has been nicknamed the "Hershey-Chase Experiment". At that time, people knew that viruses were composed of DNA (or RNA) inside a protein coat/shell called a capsid. It was also known that viruses replicate by taking over the host cell's metabolic functions to make more virus. We are used to thinking and talking about viruses which invade our bodies and make us sick, but there are other, different kinds of viruses that infect other kinds of animals, still other viruses which infect plants, and even some viruses that infect bacteria.

A virus which infects a bacterium is called a bacteriophage because the host bacterium cell is killed as the new virus particles leave the bacterial cell. In order to do all this, the virus must inject whatever is the viral genetic code into the host cell.

Thus, people realised that the viral genetic code material had to be either its DNA or its protein capsid. Hershey and Chase sought an answer to the question, "Is it the viral DNA or viral protein coat (capsid) that is the viral genetic code material which gets injected into a host bacterium cell? To try to answer this question, Hershey and Chase performed an experiment using a bacterium named *Escherichia coli*, or *E. coli* for short (named after a scientist whose last name was Escher) and a virus called T2 that is a bacteriophage that infects *E. coli*. Isolated T2, like other viruses, is just a crystal of DNA and protein, so it must live inside *E. coli* in order to make more virus like itself.

When the new T2 viruses are ready to leave the host *E. coli* cell (and go infect others), they burst the *E. coli* cell open, killing it (hence the name "bacteriophage"). The results that Hershey and Chase obtained indicated that the viral DNA, not the protein, is its genetic code material. Hershey and Chase used radioactive chemicals to distinguish between ("label") the protein capsid and the DNA in T2 virus so they could tell which of those molecules entered the *E. coli* cells. Since some amino acids contain sulpher in their side chains, if T2 is grown in *E. coli* with a source of radioactive sulpher, the sulpher will be incorporated into the T2 protein coat making it radioactive. Since DNA has lots of phosphorus in its phosphate ($-PO_4$) groups, if T2 is grown in *E. coli* with a source of radioactive phosphorus, the phosphorus will be incorporated into the viral DNA, making that radioactive.

Hershey and Chase grew two batches of T2 and *E. coli*: one with radioactive sulpher and one with radioactive phosphorus to get batches of T2 "labeled" with either radioactive S or radioactive P. Then, these radioactive T2 were placed in separate, new batches of *E. coli*, but were left there only 10 minutes. This was to give the T2 time to inject their genetic material into the bacteria, but not reproduce. In the next step, still in separate batches, the mixtures were agitated in a kitchen blender to knock loose any viral parts not inside the *E. coli* but perhaps stuck on the outer surface.

Hopefully, this would differentiate between the protein and DNA portions of the virus. Then, each mixture was spun in a centrifuge to separate the heavy bacteria (with any viral parts that had gone into them) from the liquid solution they were in (including any viral parts that had not entered the bacteria). The centrifuge causes the heavier bacteria to be pulled to the bottom of the tube where they form a pellet, while the light-weight viral "left-overs" stay suspended in the liquid portion called the supernatant.

In the subsequent step, the pellet and supernatant from each tube were separated and tested for the presence of radioactivity. Radioactive sufur was found in the supernatant, indicating that the viral protein did not go into the bacteria. Radioactive phosphorus was found in the bacterial pellet, indicating that viral DNA did go into the bacteria.

Based on these results, Hershey and Chase concluded that DNA must be the genetic code material, not protein as many poeple believed. When their experiment was published and people finally acknowledged that DNA was the genetic material, there was a lot of competition to be the first to discover its chemical structure. What was known is that DNA contains a nitrogenous base.

DNA is a Double Helix

The outer edges are formed of alternating ribose sugar molecules and phosphate groups. The two strands go in opposite directions (1 "up" and 1 "down"). The nitrogenous bases are "inside" like rungs on a ladder. Adenine on one side pairs with thymine (uracil in RNA) on the other by hydrogen bonding, and cytosine pairs with guanine. Note that the C-G pair has three hydrogen bonds while the A-T pair has only two, which keeps them from pairing wrong.

This dictates side-to-side pairing, but says nothing about the order *along* the molecule. Watson and Crick said this variability along the molecule can account for the variety in the genetic code. Their model also accounts for how DNA can replicate itself. They said the molecule "unzips" and new matching bases are added in to create two new molecules. They called this semiconservative replication because each new molecule has one "old" and one "new" strand of DNA.

DNA codes for protein synthesis by first coding for RNA. First, the DNA code is transcribed to RNA code, which is still in the "language" of nitrogenous bases, except that adenine on the DNA pairs with uracil (in place of thymine) on the RNA. The RNA code is then translated to protein code, which is a different "language."

This process involves ribosomes and two kinds of RNA: mRNA and tRNA. The mRNA codes for the gene in question and is copied off the DNA, while tRNA matches a specific group of nucleotides with a specific amino acid. A "unit" of three nucleotides on the tRNA codes for one amino acid. Each of these "units" is called an anticodon. These match up with corresponding three-nucleotide sequences on the mRNA called codons, and in this manner the amino acids are organized into the correct sequence to build a protein. The ribosome works with the mRNA and tRNA to hook the amino acids together to form a protein. Here is a list of the mRNA codons and the corresponding amino acids for which they code. Mutations can be caused by a change in the sequence of the nucleotides. Some mutations have more effect than others, depending on where in the code they are and how important that area is to the code. While mutations in some areas of some genes have little effect, sickle cell anemia is caused by a mutation in only one nucleotide. This changes the codon at that location to code for a different amino acid, and that, in turn, significantly changes the shape of the hemoglobin molecules in that person's blood. When some viruses (especially *Herpes* viruses, including Chicken Pox and Cold Sores) infect us, they insert their DNA into our cells' DNA, and stay resident in our cells for the rest of our lives. These can potentially become active again either making a person sick again or just being shed from a person's body (to infect others) without obvious symptoms of illness (like Mononucleosis). Some kinds of cancer may be caused this way. For example, there is some pretty strong evidence linking genital warts (human papillomavirus, HPV) and cervical cancer.

The AIDS virus does things "backwards." This virus contains RNA rather than DNA, yet when it gets into someone's cells, it can do reverse transcription and code from its RNA to make DNA which, then, can code to make more virus.

Genetic Engineering

We now have the knowledge and ability to transfer genes from one organism to another, which seems to have some benefits associated with it, but may also have many yet-to-be-discovered problems associated with it. Because this is all so new, not enough time has elapsed to allow scientists to study/look for any possible long-term effects of genetically-modified organisms (GMOs). Many medicines are now made by GMOs. For example, insulin was formerly extracted from the pancreas of animals after they were slaughtered for meat. However, now most insulin is produced by bacteria with the insulin gene spliced onto their chromosome. Using this method of production, drug companies can make more insulin, faster.

In theory, insulin produced in this way should be more "pure" — someone who could not use pig insulin due to a pork allergy may be able to tolerate insulin made in this way (but someone could, potentially, be allergic to some component of the bacteria present in the refined insulin). In the relatively short time that this form of insulin has been available, there is no doubt that it has saved many people's lives — let's hope that some unforseen, long-term, deleterious effects are not discovered later on.

Experimentation is being done to investigate the possible use of genetically-engineered viruses to treat genetic diseases such as cystic fibrosis. In this "treatment," a kind of virus that infects our lungs is used. The genes that enable it to infect our cells are kept while the genes which make us sick are (hopefully) all removed, and the missing human (normal) gene that relieves cystic fibrosis is then inserted into the virus' genome.

These viruses are then sprayed into the lungs of a person with CF and allowed to "infect" the cells in that person's lungs. When the normal gene is inserted into the genetic make-up of the cells lining that person's lungs, those cells function normally and the CF symptoms are alleviated. However, this treatment doesn't last because only that layer of cells is "infected," and when those cells die and are replaced by new cells, the new cells do not contain the genetic code to overcome the CF gene, and the person must inhale more genetically-engineered virus. While this seems to be a promising, life-giving, technique, no data are yet available on long-term effects and safety.

Bacillus Thuringiensis (BT)

Bacillus thuringiensis (BT) is a species of bacterium that infects and kills a number of species of caterpillars (Remember that caterpillars turn into butterflies or moths when they grow up.). There is a species of moth whose caterpillar is a "pest" in corn plants, and insects of any kind are more successful when we humans plant huge monocultures of their favourite foods. For a number of years, now, people have realised that in certain situations, by judiciously applying BT, "pest" species of caterpillars can be infected and killed without the use of man-made, chemical insecticides.

More recently, a major US chemical company came up with the idea of creating genetically-engineered corn containing BT genes, which was good news to agri-business firms who plant huge areas of land with monocultures of corn. Because corn is wind-pollinated and therefore makes lots of pollen which, in this case, contains BT genes, many scientists are concerned about the effects of this corn on local butterfly populations. Some research has indicated that when this pollen settles on nearby caterpillar host plants and is, therefore, consumed by caterpillars, this might cause an increase in mortality (therefore fewer "good" butterflies such as Monarchs). It is assumed that these BT genes in corn should, they think, have no adverse effects on people or cattle who consume this corn, but no long-term testing has been done. Also, this company has convinced the government that this corn should be marketed without any labeling indicating that it is genetically-engineered — they're scared that if we are given a choice, we won't buy/eat their corn if we know it is genetically engineered.

Additionally, to increase their profits, this company has also put "suicide" genes into this corn so that farmers cannot save seed from one year to plant the next year, and have to buy more seed from them, instead. For big agri-businesses, this is of small consequence, but for small, family farmers this is total disaster. To save money, the latter often save seed from one year to plant the next, and even if they don't plant this genetically-engineered corn, if their corn is pollinated by genetically-engineered corn from a neighboring field, it will not produce viable seed.

This same chemical company came up with the idea of "transplanting" what they think is the cold-tolerance gene from a species of cold-water-inhabiting fish into tomatoes, thereby hoping to "invent" cold-tolerant tomatoes, and again, has convinced the government that these tomatoes should not be labeled in any way to indicate that they have fish genes in them, and that we should not have a choice about what we eat. At the very least, this would be a problem for someone who is a vegetarian and chooses to not eat fish. A more serious consequence, however, would be that a person who is severely allergic to fish could also have an allergic reaction to these tomatoes and end up in the hospital. Again, no tests were done before the government approved these tomatoes and no data are available on the long-term effects on humans (or anything else).

This same chemical company manufactures a widely-used herbicide, and came up with the idea to genetically engineer cotton, soybeans, and other crop plants so they would be immune to the effects of that herbicide. That way, farmers can buy their seed from that company, then spray their fields with herbicide also purchased from that company to (hopefully) kill all local plants except the immune crop, simultaneously putting all their money in the corporation's pockets. Preliminary research has shown that these resistant genes have already begun to "jump" into local weeds, thereby making them resistant to that herbicide, and again, no data are available on long-term effects on humans, other animals, or the environment in general.

Structure of DNA

Most biological experiments are done with samples of living matter- cells, tissues, whole organisms or extracts of these materials. Only a few experiments are done with mathematical or physical models of biological components. Ocassionally, however, a model experiment gives an insight that would be difficult to obtain in any other way. This was true of the discovery of the structure of DNA. A crucial experiment was done by James Watson in 1952 using nothing more complicated than chemical models cut out of cardboard.

Inheritance

While the period from the early 1900s to World War II has been considered the "golden age" of genetics, scientists still had not determined that DNA, and not protein, was the hereditary material. However, during this time a great many genetic discoveries were made and the link between genetics and evolution was made. Friedrich Meischer in 1869 isolated DNA from fish sperm and the pus of open wounds. Since it came from nuclei, Meischer named this new chemical, nuclein. Subsequently the name was changed to nucleic acid and lastly to deoxyribonucleic acid (DNA).

Robert Feulgen, in 1914, discovered that fuchsin dye stained DNA. DNA was then found in the nucleus of all eukaryotic cells. During the 1920s, biochemist P.A. Levene analysed the components of the DNA molecule. He found it contained four nitrogenous bases: cytosine, thymine, adenine, and guanine; deoxyribose sugar; and a phosphate group. He concluded that the basic unit (nucleotide) was composed of a base attached to a sugar and that the phosphate also attached to the sugar. He (unfortunately) also erroneously concluded that the proportions of bases were equal and that there was a tetranucleotide that was the repeating structure of the molecule. The nucleotide, however, remains as the fundemantal unit (monomer) of the nucleic acid polymer. There are four nucleotides: those with cytosine (C), those with guanine (G), those with adenine (A), and those with thymine (T). During the early 1900s, the study of genetics began in earnest: the link between Mendel's work and that of cell biologists resulted in the chromosomal theory of inheritance; Garrod proposed the link between genes and "inborn errors of metabolism"; and the question was formed: what is a gene?

The answer came from the study of a deadly infectious disease: *pneumonia*. During the 1920s Frederick Griffith studied the difference between a disease-causing strain of the pneumonia causing bacteria (*Streptococcus peumoniae*) and a strain that did not cause pneumonia. The pneumonia-causing strain (the S strain) was surrounded by a capsule. The other strain (the R strain) did not have a capsule and also did not cause pneumonia. Frederick Griffith was able to induce a nonpathogenic strain of the bacterium *Streptococcus pneumoniae* to become pathogenic. Griffith referred to a transforming factor that caused the non-pathogenic bacteria to become pathogenic.

Griffith injected the different strains of bacteria into mice. The S strain killed the mice; the R strain did not. He further noted that if heat killed S strain was injected into a mouse, it did not cause pneumonia. When he combined heat-killed S with Live R and injected the mixture into a mouse (remember neither alone will kill the mouse) that the mouse developed pneumonia and died. Bacteria recovered from the mouse had a capsule and killed other mice when injected into them!

Hypotheses:

- The dead S strain had been reanimated/resurrected.
- The Live R had been transformed into Live S by some "transforming factor".

Further experiments led Griffith to conclude that number 2 was correct. In 1944, Oswald Avery, Colin MacLeod, and Maclyn McCarty revisited Griffith's experiment and concluded the transforming factor was DNA. Their evidence was strong but not totally conclusive. The then-current favourite for the hereditary material was protein; DNA was not considered by many scientists to be a strong candidate.

The breakthrough in the quest to determine the hereditary material came from the work of Max Delbruck and Salvador Luria in the 1940s. Bacteriophage are a type of virus that attacks bacteria, the viruses that Delbruck and Luria worked with were those attacking *Escherichia coli*, a bacterium found in human intestines. Bacteriophages consist of protein coats covering DNA. Bacteriophages infect a cell by injecting DNA into the host cell. This viral DNA then "disappears" while taking over the bacterial machinery and beginning to make new virus instead of new bacteria. After 25 minutes the host cell bursts, releasing hundres of new bacteriophage.

Phages have DNA and protein, making them ideal to resolve the nature of the hereditary material. In 1952, Alfred D. Hershey and Martha Chase conducted a series of experiments to determine whether protein or DNA was the hereditary material. By labeling the DNA and protein with different (and mutually exclusive) radioisotopes, they would be able to determine which chemical (DNA or protein) was getting into the bacteria.

Such material must be the hereditary material (Griffith's transforming agent). Since DNA contains Phosphorous (P) but no Sulpher (S), they tagged the DNA with radioactive Phosphorous-32. Conversely, protein lacks P but does have S, thus it could be tagged with radioactive Sulpher-35. Hershey and Chase found that the radioactive S remained outside the cell while the radioactive P was found inside the cell, indicating that DNA was the physical carrier of heredity.

Erwin Chargaff analysed the nitrogenous bases in many different forms of life, concluding that the amount of purines does not always equal the amount of pyrimidines (as proposed by Levene). DNA had been proven as the genetic material by the Hershey-Chase experiments, but how DNA served as genes was not yet certain. DNA must carry information from parent cell to daughter cell. It must contain information for replicating itself. It must be chemically stable, relatively unchanging. However, it must be capable of mutational change. Without mutations there would be no process of evolution.

Many scientists were interested in deciphering the structure of DNA, among them were Francis Crick, James Watson, Rosalind Franklin, and Maurice Wilkens. Watson and Crick gathered all available data in an attempt to develop a model of DNA structure. Franklin took X-ray diffraction photomicrographs of crystalline DNA extract, the key to the puzzle. The data known at the time was that DNA was a long molecule, proteins were helically coiled (as determined by the work of Linus Pauling), Chargaff's base data, and the x-ray diffraction data of Franklin and Wilkens.

DNA is a double helix, with bases to the centre (like rungs on a ladder) and sugar-phosphate units along the sides of the helix (like the sides of a twisted ladder). The strands are complementary (deduced by Watson and Crick from Chargaff's data, A pairs with T and C pairs with G, the pairs held together by hydrogen bonds). Notice that a double-ringed purine is always bonded to a single ring pyrimidine. Purines are Adenine (A) and Guanine (G). We have encountered Adenosine triphosphate (ATP) before, although in that case the sugar was ribose, whereas in DNA it is deoxyribose. Pyrimidines are Cytosine (C) and Thymine (T). The bases are complementary, with A on one side of the molecule we only get T on the other side, similarly with G and C. If we know the base sequence of one strand we know its complement.

Hereditary Material

DNA was proven as the hereditary material and Watson et al. had deciphered its structure. What remained was to determine how DNA copied its information and how that was expressed in the phenotype. Matthew Meselson and Franklin W. Stahl designed an experiment to determine the method of DNA replication. Three models of replication were considered likely.

Dispersive Replication: Involved the breaking of the parental strands during replication, and somehow, a reassembly of molecules that were a mix of old and new fragments on each strand of DNA. The Meselson-Stahl experiment involved the growth of *E. coli* bacteria on a growth medium containing heavy nitrogen (Nitrogen-15 as opposed to the more common, but lighter molecular weight isotope, Nitrogen-14). The first generation of bacteria was grown on a medium where the sole source of N was Nitrogen-15. The bacteria were then transferred to a medium with light (Nitrogen-14) medium. Watson and Crick had predicted that DNA replication was semi-conservative. If it was, then the DNA produced by bacteria grown on light medium would be intermediate between heavy and light.

It was. DNA replication involves a great many building blocks, enzymes and a great deal of ATP energy (remember that after the S phase of the cell cycle cells have a G phase to regenerate energy for cell division). Only occurring in a cell once per (cell) generation, DNA replication in humans occurs at a rate of 50 nucleotides per second, 500/second in prokaryotes. Nucleotides have to be assembled and available in the nucleus, along with energy to make bonds between nucleotides.

DNA polymerases unzip the helix by breaking the H-bonds between bases. Once the polymerases have opened the molecule, an area known as the replication bubble forms (always initiated at a certain set of nucleotides, the origin of replication). New nucleotides are placed in the fork and link to the corresponding parental nucleotide already there (A with T, C with G). Prokaryotes open a single replication bubble, while eukaryotes have multiple bubbles. The entire length of the DNA molecule is replicated as the bubbles meet. Since the DNA strands are antiparallel, and replication proceeds in thje 5' to 3' direction on EACH strand, one strand will form a continuous copy, while the other will form a series of short Okazaki fragments.

Enzyme Structure of DNA

Before a cell can begin to divide or differentiate, the genetic information within the cell's DNA must be copied, or "transcribed," onto complementary strands of RNA. RNA polymerase II (pol II) is an enzyme that, by itself, can unwind the DNA double helix, synthesize RNA, and proofread the result. When combined with other molecules that regulate and control the transcription process, pol II is the key to successful interpretion of an organism's genetic code.

However, the size, complexity, scarcity, and fragility of pol II complexes have made analysis of these macromolecules by x-ray crystallography a formidable challenge. A team of structural biologists has met this challenge using data obtained from both the Stanford Synchrotron Radiation Laboratory and the Macromolecular Crystallography Facility at the ALS. The resultant high-resolution model of a 10-subunit pol II complex suggests roles for each of the subunits and will allow researchers to begin unraveling the intricacies of DNA transcription and its role in gene expression.

In this work, the researchers studied the pol II enzyme from the yeast *Saccharomyces cerevisiae*, which is likely to be an excellent model for the human enzyme in light of its highly similar gene sequences. It is also the best-characterized form of the pol II enzyme, having been the subject of many biochemical and low-resolution structural studies in the past. To obtain a high-resolution structure, the research team drew on its considerable expertise in the preparation of protein crystals: two-dimensional crystals of pol II (minus two small subunits found to impede crystal growth) were used as seeds for growing three-dimensional crystals. These crystals, when produced in an inert atmosphere to prevent oxidation, enabled the collection of data to 3.5-angstrom resolution. The addition of a final soaking procedure to produce uniform crystals, combined with high-brightness x-ray sources, resulted in a resolution of 3.0 angstroms. The current results bring into focus the somewhat fuzzy features previously observed in or inferred from earlier experiments.

More importantly, the structural details suggest possible explanations for some of the unusual characteristics of this enzyme, which include a high processivity (the ability to synthesize very long strands of RNA) and the tendency to work in periodic spurts separated by pauses. While it is known that additional proteins (transcription factors) play a role in controlling the activity of pol II, scientists have yet to understand how such proteins interact with pol II binding sites to perform their various functions. The pol II model reported here establishes the positions of the various subunits and provides detailed information about the DNA/RNA binding domains. The data reveal two main subunits separated by a deep cleft where DNA can enter the complex. At the end of the cleft is the active site, where the DNA can be unwound for a short distance (the "transcription bubble") and a DNA/RNA hybrid can be produced. Two prominent grooves lead away from the active site, either of which could accommodate the exiting RNA transcript. An opening below the active site may allow the entry of nucleotides (for manufacturing RNA) and transcription factors (for regulating the process).

The same opening may provide room for the leading end of the RNA strand during "backtracking" maneuvers, which are important for proofreading and for traversing obstacles such as DNA damage. Other notable features that might help account for the great stability of this transcribing complex include a pair of "jaws" that appear to grip the DNA strands as they enter the complex and, closer to the active site, a clamp on the DNA that could possibly be locked in the closed position by the presence of RNA. The high-resolution pol II structure reported here is a landmark achievement, pulling together threads from numerous diverse research efforts into a cohesive whole. Further study should yield many new insights into the detailed mechanisms of pol II and its transcription factors. Construction of an atomic model is already well underway.

Unraveling DNA

Encoded into the double-helical strands of DNA, the human genome is the complete set of instructions required to make a human being. While the mapping of the human genome (roughly three billion components) is certainly a Herculean accomplishment, it is only the first step towards realizing the full potential of genomic medicine in the diagnosis, monitoring, and treatment of disease. Beyond knowing what the genetic blueprint says, scientists must understand how that blueprint gets interpreted, or "expressed" as an individual with unique traits. The pol II enzyme is the catalyst for a major step in this process.

As a pol II molecule slides along a DNA molecule, it "unzips" the strands of the DNA double helix, synthesizes a complementary strand of RNA (which will carry the genetic information to where it is needed), and verifies that no mistakes have occurred. This process is regulated by transcription factors—separate molecules that bind to pol II and determine which genes are expressed, at what stage of development, and in which tissue.

Done correctly, this process results in healthy cell growth and differentiation; otherwise, aberrations such as cancer can be the result. Thus, details of the structure of pol II, including information about its binding sites and how they interact with transcription factors, will provide valuable insight into the detailed mechanisms underlying the flow of genetic information from DNA to RNA to protein, which is necessary for life and health.

Without hydrogen bonds there could be no life because they hold the double helix of DNA together, and this they do by charge attractions. A hydrogen bonded to oxygen, or nitrogen, becomes slightly positively charged which enables it to attract a centre of negative charge on another molecule, such as another oxygen or nitrogen atom. The hydrogen bond is then written, *e.g.*, O–H...N, with the dotted line signifying the hydrogen bond. There are also O–H...O, N–H...O and N–H...N bonds, the last being among the weakest.

The secondary effects they have on structures, molecular vibrations, etc., can be used to infer hydrogen bonding, but there is no primary way of observing them because of their inherent weakness. NMR appears to be the least useful technique because neither of the common isotopes, oxygen-16 or nitrogen-14, has a magnetic nucleus. However, nitrogen-15 has a magnetic moment, and by replacing ^{14}N by N, Grzesiek has opened up a new area of investigating these enigmatic bonds. Paper reports for the first time the direct observation by NMR of an N–H...N hydrogen bond between nucleic acids enriched with ^{15}N, by measuring the coupling of the nitrogen atoms. Grzesiek, working with Andrew Dingley of the Heinrich-Heine University in Düsseldorf, has been able to do this and show that the coupling is surprisingly large. Normally atomic nuclei only couple with each other if they are linked by normal chemical bonds, and in theory hydrogen bonds have neither the strength nor stability for this to occur.

The German researchers studied an ^{15}N enriched sample of the T1 domain of the potato spindle tuber viroid and were able to prove that N–H...N hydrogen bonding was present between the base pairs, uridine...adenosine and guanosine...cytidine, with couplings of approximately 7 Hz. How could they be certain that the signals they were observing are due to N–H...N hydrogen bonds?

The answer ws to use triple resonance techniques to examine base pairs that hydrogen bond only via O–H...N hydrogen bonds, and show that the signal they had previously observed was absent. Grzesiek's second paper on hydrogen bonds, coauthored by Florence Cordier, Heinrich-Heine University, extends the work in an even more remarkable way by measuring the NMR coupling between nitrogens and carbons in the backbone hydrogen bonds of the human protein ubiquitin. The carbons are part of a carbonyl (C=O) group, so are one removed from the hydrogen bond, *i.e.*, N–H...O=C. This time they used material enriched with ^{15}N and ^{13}C (normal ^{12}C has no nuclear magnet), and there too was the evidence for these hydrogen bonds, albeit with an interaction an order of magnitude weaker (at –0.25 to –0.9 Hz) than the N–H...N coupling. Nevertheless, the couplings correlate with the strength of the hydrogen bond, being stronger in the stronger bonds. Grzesiek's third paper, was done in conjunction with Dingley and researchers at UCLA. Together they studied not only the hydrogen bonding of Watson-Crick base pairs but also of Hoogsten base pairs within a DNA triplex consisting of one purine and two pyrimidine strands.

Four different base pairs were identified, their various couplings distinguished–including those of the weaker interactions at the "frayed ends" of the DNA chains–and relationships with other hydrogen bonding parameters, such as bond length, were established. In addition they were able to show that density functional computer simulations by computer could reproduce these findings exactly.

Ribonucleic Acid

Ribonucleic acid (RNA) is one of the three major macromolecules (along with DNA and proteins) that are essential for all known forms of life. Like

DNA, RNA is made up of a long chain of components called nucleotides. Each nucleotide consists of a nucleobase (sometimes called a nitrogenous base), a ribose sugar, and a phosphate group. The sequence of nucleotides allows RNA to encode genetic information. For example, some viruses use RNA instead of DNA as their genetic material, and all organisms use messenger RNA (mRNA) to carry the genetic information that directs the synthesis of proteins.

Like proteins, some RNA molecules play an active role in cells by catalyzing biological reactions, controlling gene expression, or sensing and communicating responses to cellular signals. One of these active processes is protein synthesis, a universal function whereby mRNA molecules direct the assembly of proteins on ribosomes. This process uses transfer RNA (tRNA) molecules to deliver amino acids to the ribosome, where ribosomal RNA (rRNA) links amino acids together to form proteins.

The chemical structure of RNA is very similar to that of DNA, with two differences:

1. RNA contains the sugar ribose while DNA contains the slightly different sugar deoxyribose (a type of ribose that lacks one oxygen atom),
2. RNA has the nucleobase uracil while DNA contains thymine (uracil and thymine have similar base-pairing properties).

Unlike DNA, most RNA molecules are single-stranded. Single-stranded RNA molecules adopt very complex three-dimensional structures, since they are not restricted to the repetitive double-helical form of double-stranded DNA. RNA is made within living cells by RNA polymerases, enzymes which act to copy a DNA or RNA template into a new RNA strand through processes known as transcription or RNA replication, respectively.

We now turn our attention to RNA, which differs from DNA in three respects. First, the backbone of RNA contains ribose rather than 2'-deoxyribose. That is, ribose has a hydroxyl group at the 2' position. Second, RNA contains uracil in place of thymine. Uracil has the same single-ringed structure as thymine, except that it lacks the 5' methyl group. Thymine is in effect 5'methyl-uracil. Third, RNA is usually found as a single polynucleotide chain.

Except for the case of certain viruses, RNA is not the genetic material and does not need to be capable of serving as a template for its own replication. Rather, RNA functions as the intermediate, the mRNA, between the gene and the protein-synthesizing machinery. Another function of RNA is as an adaptor, the tRNA, between the codons in the mRNA and amino acids. RNA can also play a structural role as in the case of the RNA components of the ribosome.

Yet another role for RNA is as a regulatory molecule, which through sequence complementarity binds to, and interferes with the translation of, certain mRNAs. Finally, some RNAs (including one of the structural RNAs of the ribosome) are enzymes that catalyze essential reactions in the cell.

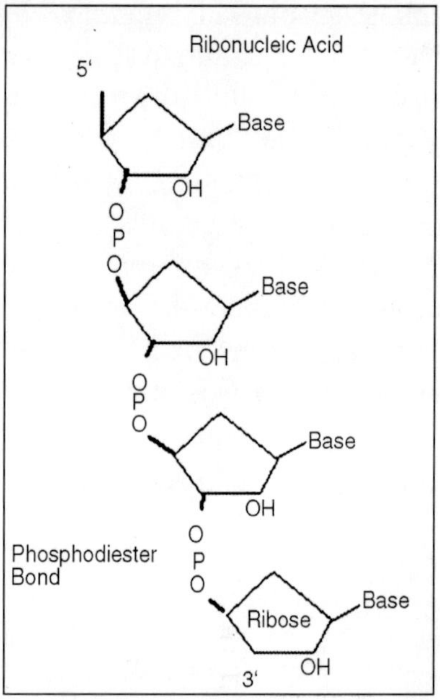

Figure: RNA Structure

In all of these cases, the RNA is copied as a single strand off only one of the two strands of the DNA template, and its complementary strand does not exist. RNA is capable of forming long double helices, but these are unusual in nature.

RNA Chains

Despite being single-stranded, RNA molecules often exhibit a great deal of double-helical character. This is because RNA chains frequently fold back on themselves to form base-paired segments between short stretches of complementary sequences. If the two stretches of complementary sequence are near each other, the RNA may adopt one of various stem-loop structures in which the intervening RNA is looped out from the end of the double-helical segment as in a hairpin, a bulge, or a simple loop.

The stability of such stem-loop structures is in some instances enhanced by the special properties of the loop. For example, a stem-loop with the "tetraloop" sequence UUCG is unexpectedly stable due to special base-stacking interactions in the loop. Base pairing can also take place between sequences that are not contiguous to form complex structures aptly named pseudoknots. The regions of base pairing in RNA can be a regular double helix or they can contain discontinuities, such as noncomplementary nucleotides that bulge out from the helix. A feature of RNA that adds to its propensity to form double-helical structures is an additional, non-Watson-Crick base pair. This is the G:U base pair, which has hydrogen bonds between N3 of uracil and the carbonyl on C6 of guanine and between the carbonyl on C2 of uracil and N1 of guanine. Because G:U base pairs can occur as well as the four conventional, Watson-Crick base pairs, RNA chains have an enhanced capacity for self-complementarity. Thus, RNA frequently exhibits local regions of base pairing but not the long-range, regular helicity of DNA.

The presence of 2'-hydroxyls in the RNA backbone prevents RNA from adopting a B-form helix. Rather, double-helical RNA resembles the A-form structure of DNA. As such, the minor groove is wide and shallow, and hence accessible, but recall that the minor groove offers little sequence-specific information. Meanwhile, the major groove is so narrow and deep that it is not very accessible to amino acid side chains from interacting proteins. Thus, the RNA double helix is quite distinct from the DNA double helix in its detailed atomic structure and less well suited for sequence-specific interactions with proteins (although some proteins do bind to RNA in a sequence-specific manner).

Complex Tertiary Structures

Freed of the constraint of forming long-range regular helices, RNA can adopt a wealth of tertiary structures. This is because RNA has enormous rotational freedom in the backbone of its non-base-paired regions. Thus, RNA can fold up into complex tertiary structures frequently involving unconventional base pairing, such as the base triples and base-backbone interactions seen in tRNAs. Proteins can assist the formation of tertiary structures by large RNA molecules, such as those found in the ribosome.

Proteins shield the negative charges of backbone phosphates, whose electrostatic repulsive forces would otherwise destabilize the structure. Researchers have taken advantage of the potential structural complexity of RNA to generate novel RNA species (not found in nature) that have specific desirable properties. By synthesizing RNA molecules with randomized sequences, it is possible to generate mixtures of oligonucleotides representing enormous sequence diversity. For example, a mixture of oligoribonucleotides of length 20 and having four possible nucleotides at each position would have a potential complexity of 420 sequences or 1012 sequences! From mixtures of diverse oligoribonucleotides, RNA molecules can be selected biochemically that have particular properties, such as an affinity for a specific small molecule.

RNAs Are Enzymes

It was widely believed for many years that only proteins could be enzymes. An enzyme must be able to bind a substrate, carry out a chemical reaction, release the product and repeat this sequence of events many times. Proteins are well suited to this task because they are composed of many different kinds of amino acids and they can fold into complex tertiary structures with binding pockets for the substrate and small molecule cofactors and an active site for catalysis.

Now we know that RNAs, which as we have seen can similarly adopt complex tertiary structures, can also be biological catalysts. Such RNA enzymes are known as ribozymes, and they exhibit many of the features of a classical enzyme, such as an active site, a binding site for a substrate and a binding site for a cofactor, such as a metal ion. One of the first ribozymes to be discovered was RNase P, a ribonuclease that is involved in generating tRNA molecules from larger, precursor RNAs. RNase P is composed of both RNA and protein; however, the RNA moiety alone is the catalyst.

The protein moiety of RNase P facilitates the reaction by shielding the negative charges on the RNA so that it can bind effectively to its negatively charged substrate. The RNA moiety is able to catalyze cleavage of the tRNA precursor in the absence of the protein if a small, positively charged counter ion, such as the peptide spermidine, is used to shield the repulsive, negative charges. Other ribozymes carry out trans-esterification reactions involved in the removal of intervening sequences known as introns from precursors to certain mRNAs, tRNAs, and ribosomal RNAs in a process known as RNA splicing.

Hammerhead Ribozyme

Before concluding our discussion of RNA, let us look in more detail at the structure and function of one particular ribozyme, the hammerhead. The hammerhead is a sequence-specific ribonuclease that is found in certain infectious RNA agents of plants known as *viroids*, which depend on self-cleavage to propagate. When the viroid replicates, it produces multiple copies of itself in one continuous RNA chain. Single viroids arise by cleavage, and this cleavage reaction is carried out by the RNA sequence around the junction.

One such self-cleaving sequence is called the *hammerhead* because of the shape of its secondary structure, which consists of three base-paired stems surrounding a core of non-complementary nucleotides required for catalysis. The tertiary structure of the ribozyme, however, looks more like a wishbone. To understand how the hammerhead works, let us first look at how RNA undergoes hydrolysis under alkaline conditions. At high pH, the 2' hydroxyl of the ribose in the RNA backbone can become deprotonated, and the resulting negatively charged oxygen can attack the scissile phosphate at the 3' position of the same ribose. This reaction breaks the RNA chain, producing a 2', 3' cyclic phosphate and a free 5' hydroxyl. Each ribose in an RNA chain can undergo this reaction, completely cleaving the parent molecule into nucleotides. Many protein ribonucleases also cleave their RNA substrates via the formation of a 2', 3' cyclic phosphate. Working at normal cellular pH, these protein enzymes use a metal ion, bound at their active site, to activate the 2' hydroxyl of the RNA. The hammerhead is a sequencespecific ribonuclease, but it too cleaves RNA via the formation of a 2', 3' cyclic phosphate.

Hammerhead-mediated cleavage involves a ribozyme-bound Mg+ ion that deprotonates the 2' hydroxyl at neutral pH, resulting in nucleophilic attack on the scissile phosphate. Because the normal reaction of the hammerhead is self-cleavage, it is not really a catalyst; each molecule normally promotes a reaction one time only, thus having a turnover number of one. But the hammerhead can be engineered to function as a true ribozyme by dividing the molecule into two portions—one, the ribozyme, that contains the catalytic core and the other, the substrate, that contains the cleavage site. The substrate binds to the ribozyme at stems I and III. After cleavage, the substrate is released and replaced by a fresh uncut substrate, thereby allowing repeated rounds of cleavage.

Life Evolve from an RNA

The discovery of ribozymes has profoundly altered our view of how life might have evolved. We can now imagine that there was a primitive form of life based entirely on RNA. In this world, RNA would have functioned as the genetic material and as the enzymatic machines. This RNA world would have preceded life as we know it today, in which information transfer is based on DNA, RNA, and protein. A hint that the protein world might have arisen from an RNA world is the discovery that the component in the ribosome that is responsible for the formation of the peptide bond, the peptidyl transferase, is an RNA molecule. Unlike RNase P, the hammerhead, and other previously known ribozymes which act on phosphorous centres, the peptidyl transferase acts on a carbon centre to create the peptide bond. It thus links RNA chemistry to the most fundamental reaction in the protein world, peptide bond formation. Perhaps then the ribosome ribozyme is a relic of an earlier form of life in which all enzymes were RNAs. DNA is usually in the form of a right-handed double helix. The helix consists of two polydeoxynucleotide chains.

Each chain is an alternating polymer of deoxyriboe sugars and phosphates that are joined together via phosphodiester linkages. One of four bases protrudes from each sugar: adenine and guanine, which are purines, and thymine and cytosine, which are pyrimidines. While the sugar phosphate backbone is regular, the order of bases is irregular and this is responsible for the information content of DNA. Each chain has a 5' to 3' polarity, and the two chains of the double helix are oriented in an antiparallel manner—that is, they run in opposite directions. Pairing between the bases holds the chains together. Pairing is mediated by hydrogen bonds and is specific: Adenine on one chain is always paired with thymine on the other chain, whereas guanine is always paired with cytosine. This strict base-pairing reflects the fixed locations of hydrogen atoms in the purine and pyrimidine bases in the forms of those bases found in DNA. Adenine and cytosine almost always exist in the amino as opposed to the imino tautomeric forms, whereas guanine and thymine almost always exist in the keto as opposed to enol forms.

The complementarity between the bases on the two strands gives DNA its self-coding character. The two strands of the double helix fall apart (denature) upon exposure to high temperature, extremes of pH, or any agent that causes the breakage of hydrogen bonds. Upon slow return to normal cellular conditions, the denatured single strands can specifically reassociate to biologically active double helices (renature or anneal).

DNA in solution has a helical periodicity of about 10.5 base pairs per turn of the helix. The stacking of base pairs upon each other creates a helix with two grooves. Because the sugars protrude from the bases at an angle of about 120°, the grooves are unequal in size. The edges of each base pair are exposed in the grooves, creating a pattern of hydrogen bond donors and acceptors and of van der Waals surfaces that identifies the base pair.

The wider—or *major*—groove is richer in chemical information than the narrow (*minor*) groove and is more important for recognition by nucleotide sequence-specific binding proteins. Almost all cellular DNAs are extremely long molecules, with only one DNA molecule within a given chromosome. Eukaryotic cells accommodate this extreme length in part by wrapping the DNA around protein particles known as nucleosomes. Most DNA molecules are linear but some DNAs are circles, as is often the case for the chromosomes of prokaryotes and for certain viruses. DNA is flexible. Unless the molecule is topologically constrained, it can freely rotate to accommodate changes in the number of times the two strands twist about each other.

DNA is topologically constrained when it is in the form of a covalently closed circle, or when it is entrained in chromatin. The linking number is an invariant topological property of covalently closed circular DNA. It is the number of times one strand would have to be passed through the other strand in order to separate the two circular strands. The linking number is the sum of two interconvertible geometric properties: twist, which is the number of times the two strands are wrapped around each other; and the writhing number, which is the number of times the long axis of the DNA crosses over itself in space. DNA is relaxed under physiological conditions when it has about 10.5 base pairs per turn and is free of writhe. If the linking number is decreased, then the DNA becomes torsionally stressed, and it is said to be negatively supercoiled. DNA in cells is usually negatively supercoiled by about 6%. The left-handed wrapping of DNA around nucleosomes introduces negative supercoiling in eukaryotes. In prokaryotes, which lack histones, the enzyme DNA gyrase is responsible for generating negative supercoils. DNA gyrase is a member of the type II family of topoisomerases. These enzymes change the linking number of DNA in steps of two by making a transient break in the double helix and passing a region of duplex DNA through the break. Some type II topoisomerases relax supercoiled DNA, whereas DNA gyrase generates negative supercoils. Type I topoisomerases also relax supercoiled DNAs but do so in steps of one in which one DNA strand is passed through a transient nick in the other strand. RNA differs from DNA in the following ways: its backbone contains ribose rather than 2'-deoxyribose; it contains the pyrimidine uracil in place of thymine; and it usually exists as a single polynucleotide chain, without a complementary chain.

As a consequence of being a single strand, RNA can fold back on itself to form short stretches of double helix between regions that are complementary to each other. RNA allows a greater range of base pairing than does DNA. Thus, as well as A:U and C:G pairing, U can also pair with G. This capacity to form a non-Watson-Crick base pair adds to the propensity of RNA to form doublehelical segments. Freed of the constraint of forming longrange regular helices, RNA can form complex tertiary structures, which are often based on unconventional interactions between bases and between bases and the sugarphosphate backbone. Some RNAs act as enzymes—they catalyze chemical reactions in the cell and in vitro. These RNA enzymes are known as ribozymes. Most ribozymes act on phosphorous centres, as in the case of the ribonuclease RNase P. RNase P is composed of protein and RNA, but it is the RNA moiety that is the catalyst. The hammerhead is a self-cleaving RNA, which cuts the RNA backbone via the formation of a 2', 3' cyclic phosphate in a reaction that involves an RNA-bound Mg^{+} ion. Peptidyl transferase is an example of a ribozyme that acts on a carbon centre. This ribozyme, which is responsible for the formation of the peptide bond, is one of the RNA components of the ribosome. The discovery of RNA enzymes that can act on phosphorous or carbon centres suggests that life might have evolved from a primitive form in which RNA functioned both as the genetic material and as the enzymatic machinery.

7

Alternate DNA Structures

DNA exists in many possible conformations that include A-DNA, B-DNA, and Z-DNA forms, although, only B-DNA and Z-DNA have been directly observed in functional organisms. The conformation that DNA adopts depends on the hydration level, DNA sequence, the amount and direction of supercoiling, chemical modifications of the bases, the type and concentration of metal ions, as well as the presence of polyamines in solution.

The first published reports of A-DNA X-ray diffraction patterns— and also B-DNA used analyses based on Patterson transforms that provided only a limited amount of structural information for oriented fibers of DNA.

An alternate analysis was then proposed by Wilkins et al., in 1953, for the in vivo B-DNA X-ray diffraction/scattering patterns of highly hydrated DNA fibers in terms of squares of Bessel functions. In the same journal, James D. Watson and Francis Crick presented their molecular modeling analysis of the DNA X-ray diffraction patterns to suggest that the structure was a double-helix.

Although the 'B-DNA form' is most common under the conditions found in cells, it is not a well-defined conformation but a family of related DNA conformations that occur at the high hydration levels present in living cells. Their corresponding X-ray diffraction and scattering patterns are characteristic of molecular paracrystals with a significant degree of disorder.

Compared to B-DNA, the A-DNA form is a wider right-handed spiral, with a shallow, wide minor groove and a narrower, deeper major groove. The A form occurs under non-physiological conditions in partially dehydrated samples of DNA, while in the cell it may be produced in hybrid pairings of DNA and RNA strands, as well as in enzyme-DNA complexes.

Segments of DNA where the bases have been chemically modified by methylation may undergo a larger change in conformation and adopt the Z form.

Here, the strands turn about the helical axis in a left-handed spiral, the opposite of the more common B form. These unusual structures can be recognized by specific Z-DNA binding proteins and may be involved in the regulation of transcription.

Alternate DNA Chemistry

For a number of years exobiologists have proposed the existence of a shadow biosphere, a postulated microbial biosphere of Earth that uses radically different biochemical and molecular processes than currently known life.

One of the proposals was the existence of lifeforms that use arsenic instead of phosphorus in DNA.

A December 2010 NASA press conference stated that the bacterium GFAJ-1, which has evolved in an arsenic-rich environment, is the first terrestrial lifeform found which may have this ability.

The bacterium was found in Mono Lake, east of Yosemite National Park. GFAJ-1 is a rod-shaped extremophile bacterium in the family Halomonadaceae that, when starved of phosphorus, may be capable of incorporating the usually poisonous element arsenic in its DNA.

This discovery may lend weight to the long-standing idea that extraterrestrial life could have a different chemical makeup from life on Earth.

The research was carried out by a team led by Felisa Wolfe-Simon, a geomicrobiologist and geobiochemist, a Postdoctoral Fellow of the NASA Astrobiology Institute with

Arizona State University. This finding has, however, faced strong criticism from the scientific community; scientists have argued that there is no evidence that arsenic is actually incorporated into biomolecules.Independent conformation of this finding has also not yet been possible.

Quadruplex Structures

DNA quadruplex formed by telomererepeats. The looped conformation of the DNA backbone is very different from the typical DNA helix.These guanine-rich sequences may stabilize chromosome ends by forming structures of stacked sets of four-base units, rather than the usual base pairs found in other DNA molecules.

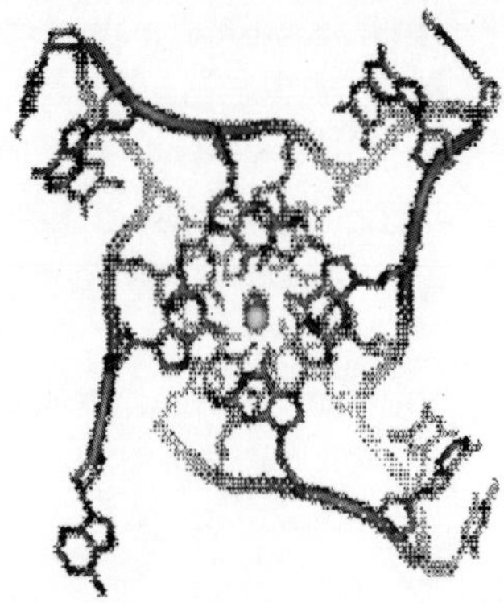

Figure: *At the ends of the linear chromosomes are specialized regions of DNA called telomeres. The main function of these regions is to allow the cell to replicate chromosome ends using the enzymetelomerase, as the enzymes that normally replicate DNA cannot copy the extreme 32 ends of chromosomes. These specialized chromosome caps also help protect the DNA ends, and stop theDNA repair systems in the cell from treating them as damage to be corrected. In human cells, telomeres are usually lengths of single-stranded DNA containing several thousand repeats of a simple TTAGGG sequence.*

Here, four guanine bases form a flat plate and these flat four-base units then stack on top of each other, to form a stable G-quadruplex structure.

These structures are stabilized by hydrogen bonding between the edges of the bases and chelation of a metal ion in the centre of each four-base unit. Other structures can also be formed, with the central set of four bases coming from either a single strand folded around the bases, or several different parallel strands, each contributing one base to the central structure.

In addition to these stacked structures, telomeres also form large loop structures called telomere loops, or T-loops. Here, the single-stranded DNA curls around in a long circle stabilized by telomere-binding proteins.

At the very end of the T-loop, the single-stranded telomere DNA is held onto a region of double-stranded DNA by the telomere strand disrupting the double-helical DNA and base pairing to one of the two strands. This triple-stranded structure is called a displacement loop or D-loop.

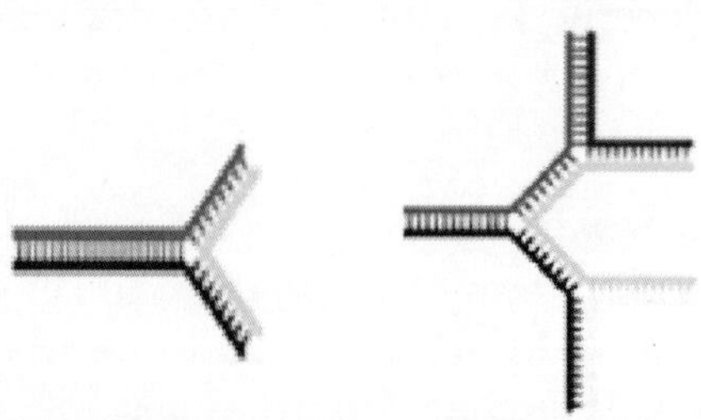

Figure: *Branched DNA can form networks containing multiple branches.*

Branched DNA

In DNA fraying occurs when non-complementary regions exist at the end of an otherwise complementary double-strand of DNA. However, branched DNA can occur if a third strand of DNA is introduced and contains adjoining regions able to hybridize with the frayed regions of the pre-existing double-strand.

Although the simplest example of branched DNA involves only three strands of DNA, complexes involving additional strands and multiple branches are also possible.Branched DNA can be used in nanotechnology to construct geometric shapes.

Vibration

DNA may carry out low-frequency collective motion as observed by the Raman spectroscopy and analyzed with a quasi-continuum model.

Chemical Modifications

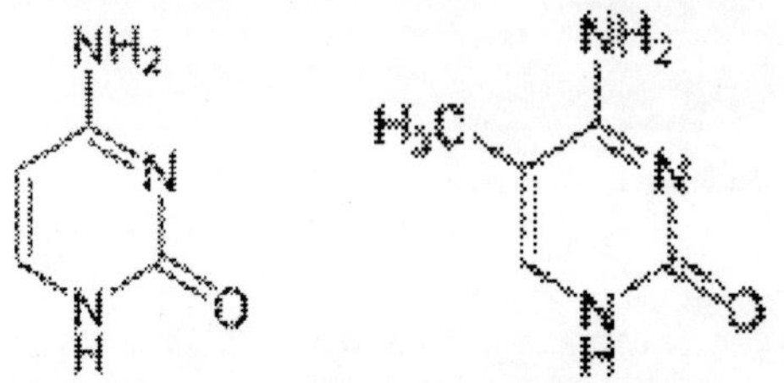

cytosine 5-methylcytosine thymine

Figure: *Structure of cytosine with and without the 5-methyl group.Deamination converts 5-methylcytosine into thymine.*

Base Modifications

The expression of genes is influenced by how the DNA is packaged in chromosomes, in a structure called chromatin. Base modifications can be involved in packaging, with regions that have low or no gene expression usually containing high levels of methylation of cytosine bases.

For example, cytosine methylation, produces 5-methylcytosine, which is important for X-chromosome inactivation. The average level of methylation varies between organisms – the worm Caenorhabditis elegans lacks cytosine methylation, while vertebrates have higher levels, with up to 1% of their DNA containing 5-methylcytosine.

Despite the importance of 5-methylcytosine, it can deaminate to leave a thymine base, so methylated cytosines are particularly prone to mutations. Other base modifications include adenine methylation in bacteria, the presence of 5-hydro-xymethylcy-tosine in the brain, and the glycosylation of uracil to produce the "J-base" in kinetoplastids.

Damage

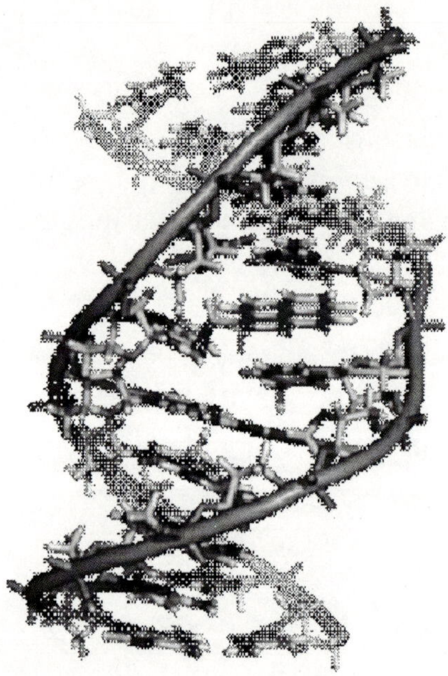

Figure: *A covalent adduct between ametabolically activated form ofbenzo[a]pyrene, the major mutagen intobacco smoke, and DNA*

DNA can be damaged by many sorts of mutagens, which change the DNA sequence. Mutagens include oxidizing agents, alkylating agents and also high-energy electromagnetic radiation such as ultraviolet light and X-rays.

The type of DNA damage produced depends on the type of mutagen. For example, UV light can damage DNA by producing thymine dimers, which are cross-links between pyrimidine bases.

On the other hand, oxidants such asfree radicals or hydrogen peroxide produce multiple forms of damage, including base modifications, particularly of guanosine, and double-strand breaks. A typical human cell contains about 150,000 bases that have suffered oxidative damage.

Of these oxidative lesions, the most dangerous are double-strand breaks, as these are difficult to repair and can produce point mutations, insertions and deletions from the DNA sequence, as well as chromosomal translocations. Many mutagens fit into the space between two adjacent base pairs, this is called intercalation. Most intercalators are aromatic and planar molecules; examples include ethidium bromide, acridines, daunomycin, and doxorubicin. In order for an intercalator to fit between base pairs, the bases must separate, distorting the DNA strands by unwinding of the double helix. This inhibits both transcription and DNA replication, causing toxicity and mutations. As a result, DNA intercalators may be carcinogens, and in the case of thalidomide, a teratogen. Others such as benzo[a]pyrene diol epoxide and aflatoxin form DNA adducts which induce errors in replication.

Nevertheless, due to their ability to inhibit DNA transcription and replication, other similar toxins are also used in chemotherapy to inhibit rapidly growing cancer cells.

Biological Functions

DNA usually occurs as linear chromosomes in eukaryotes, and circular chromosomes in prokaryotes.

The set of chromosomes in a cell makes up itsgenome; the human genome has approximately 3 billion base pairs of DNA arranged into 46 chromosomes.

The information carried by DNA is held in thesequence of pieces of DNA called genes. Transmission of genetic information in genes is achieved via complementary base pairing. ‘

For example, in transcription, when a cell uses the information in a gene, the DNA sequence is copied into a complementary RNA sequence through the attraction between the DNA and the correct RNA nucleotides. Usually, this RNA copy is then used to make a matching protein sequence in a process called translation, which depends on the same interaction between RNA nucleotides.

In alternative fashion, a cell may simply copy its genetic information in a process called DNA replication. The details of these functions are covered in other articles; here we focus on the interactions between DNA and other molecules that mediate the function of the genome.

Genes and Enomes

Genomic DNA is tightly and orderl packed in the process called DNA condensation to fit the small available volumes of the cell. In eukaryotes, DNA is locatd in the cell nucleus, as well as small amounts in mitochondria and chloroplasts. In prokaryotes, the DNA is held within an irregularly shaped body in the cytoplasm called the nucleoid.

The genetic information in a genome is held within genes, and the complete set of this information in an organism is called its genotype.

A gene is a unit of heredity and is a region of DNA that influences a particular characteristic in an organism. Genes contain an open reading frame that can be transcribed, as well as regulatory sequences such as promoters and enhancers, which control the transcription of the open reading frame.

In many species, only a small fraction of the total sequence of the genome encodes protein.

For example, only about 1.5% of the human genome consists of protein-coding exons, with over 50% of human DNA consisting of non-coding repetitive sequences.

The reasons for the presence of so much noncoding DNA in eukaryotic genomes and the extraordinary differences in genome size, or C-value, among species represent a long-standing puzzle known as the "C-value enigma".

However, DNA sequences that do not code protein may still encode functional non-coding RNA molecules, which are involved in the regulation of gene expression.

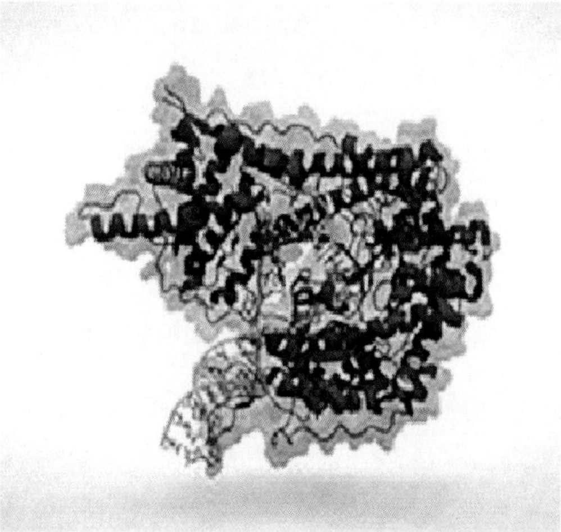

Figure: *T7 RNA polymerase (blue) producing a mRNA (green) from a DNA template (orange).*

Some noncoding DNA sequences play structural roles in chromosomes. Telomeres and centromeres typically contain few genes, but are important for the function and stability of chromosomes. An abundant form of noncoding DNA in humans are pseudogenes, which are copies of genes that have been disabled by mutation. These sequences are usually just molecular fossils, although they can occasionally serve as raw genetic material for the creation of new genes through the process of gene duplication and divergence.

Transcription and Translation

A gene is a sequence of DNA that contains genetic information and can influence the phenotype of an organism. Within a gene, the sequence of bases along a DNA strand defines a messenger RNA sequence, which then defines one or more protein sequences. The relationship between the nucleotide sequences of genes and the amino-acid sequences of proteins is determined by the rules of translation, known collectively as the genetic code. The genetic code consists of three-letter 'words' called codons formed from a sequence of three nuleotides (e.g. ACT, CAG, TTT).

In transcription, the codons of a gene are copied into messenger RNA by RNA polymerase. Tis RNA copy is then decoded by a ribosome that reads the RNA sequence by base-pairing the messenger RNA to transfer RNA, which carries amino acids.

Since there are 4 bases in 3-letter combinations, there are 64 possible codons (4 combinations).

These encode the twenty standard amino acids, giving most amino acids more than one possible codon. There are also three 'stop' or 'nonsense' codons signifying the end of the coding region; these are the TAA, TGA and TAG codons.

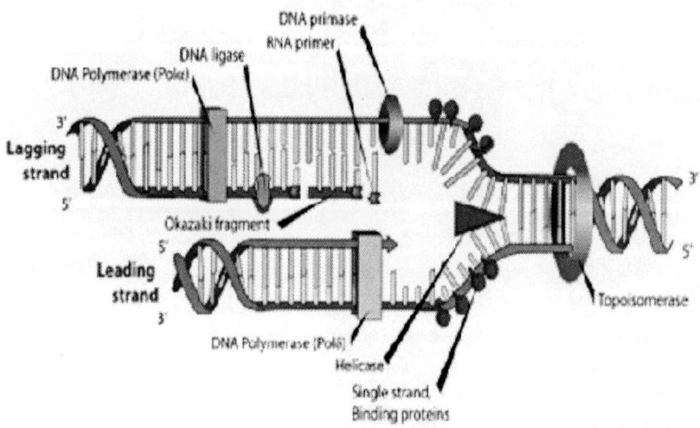

Figure: *DNA replication. The double helix is unwound by a helicase and topoisomerase. Next, one DNA polymerase produces the leading strand copy. Another DNA polymerase binds to the lagging strand. This enzyme makes discontinuous segments (called Okazaki fragments) before DNA ligase joins them together.*

Replication

Cell division is essential for an organism to grow, but, when a cell divides, it must replicate the DNA in its genome so that the two daughter cells have the same genetic information as their parent. The double-stranded structure of DNA provides a simple mechanism for DNA replication.

Here, the two strands are separated and then each strand'scomplementary DNA sequence is recreated by an enzyme called DNA polymerase. This enzyme makes the complementary strand by finding the correct base through complementary base pairing, and bonding it onto the original strand. As DNA polymerases can only extend a DNA strand in a 52 to 32 direction, different mechanisms are used to copy the antiparallel strands of the double helix. In this way, the base on the old strand dictates which base appears on the new strand, and the cell ends up with a perfect copy of its DNA.

Interactions with proteins

All the functions of DNA depend on interactions with proteins. These protein interactions can be non-specific, or the protein can bind specifically to a single DNA sequence. Enzymes can also bind to DNA and of these, the polymerases that copy the DNA base sequence in transcription and DNA replication are particularly important.

DNA-binding proteins

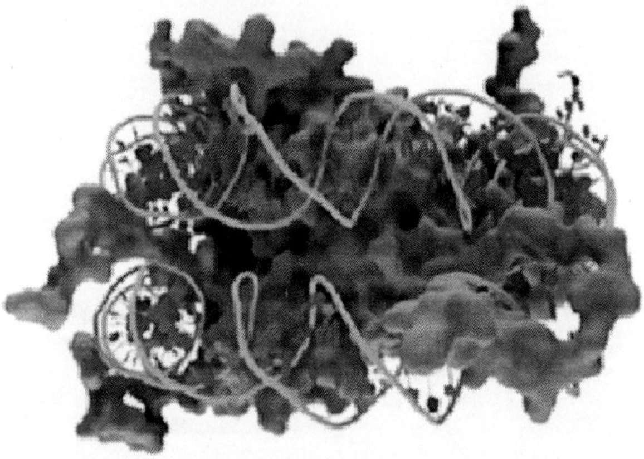

Figure: *Interaction of DNA (shown in orange) withhistones (shown in blue). These proteins' basic amino acids bind to the acidic phosphate groups on DNA.*

Structural proteins that bind DNA are well-understood examples of non-specific DNA-protein interactions. Within chromosomes, DNA is held in complexes with structural proteins.

These proteins organize the DNA into a compact structure called chromatin. In eukaryotes this structure involves DNA binding to a complex of small basic proteins called histones, while in prokaryotes multiple types of proteins are involved. The histones form a disk-shaped complex called a nucleosome, which contains two complete turns of double-stranded DNA wrapped around its surface. These non-specific interactions are formed through basic residues in the histones making ionic bonds to the acidic sugar-phosphate backbone of the DNA, and are therefore largely independent of the base sequence.

Chemical modifications of these basic amino acid residues include methylation, phosphorylationand acetylation. These chemical changes alter the strength of the interaction between the DNA and the histones, making the DNA more or less accessible to transcription factors and changing the rate of transcription. Other non-specific DNA-binding proteins in hromatin include the high-mobility group proteins, which bind to bent or distorted DNA. These proteins are important in bending arrays of nucleosomes and arranging them into the larger structures that make up chromosomes.

A distinct group of DNA-binding proteins are the DNA-binding proteins that specifically bind single-stranded DNA. In humans, replication protein A is the best-understood member of this family and is used in processes where the double helix is separated, including DNA replication, recombination and DNA repair. These binding proteins seem to stabilize single-stranded DNA and protect it from forming stem-loops or being degraded by nucleases.

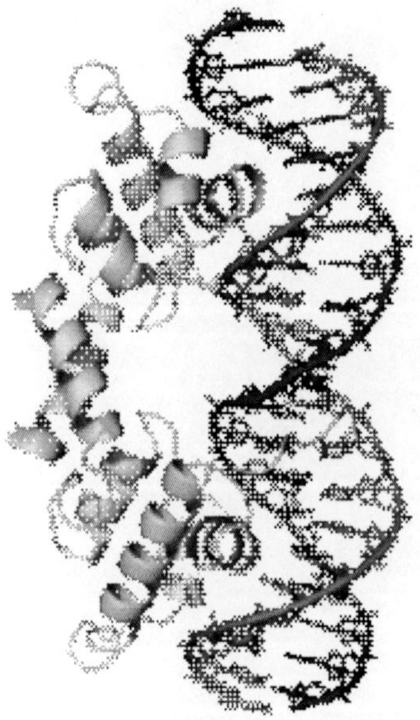

Figure: *The lambda repressor helix-turn-helix transcription factor bound to its DNA target*

In contrast, other proteins have evolved to bind to particular DNA sequences. The most intensively studied of these are the various transcription factors, which are proteins that regulate transcription. Each transcription factor binds to one particular set of DNA sequences and activates or inhibits the transcription of genes that have these sequences close to their promoters.

The transcription factors do this in two ways. Firstly, they can bind the RNA polymerase esponsible for transcription, either directly or through other mediator proteins; this locates the polymerase at the promoter and allows it to begin transcription. Alternatively, transcription factors can bind enzymes that modify the histones at the promoter; this will change the accessibility of the DNA template to the polymerase.

As these DNA targets can occur throughout an organism's genome, changes in the activity of one type of transcription factor can affect thousands of genes.Consequently, these proteins are often the targets of the signal transduction processes that control responses to environmental changes or cellular differentiation and development. The specificity of these transcription factors' interactions with DNA come from the proteins making multiple contacts to the edges of the DNA bases, allowing them to "read" the DNA sequence. Most of these base-interactions are made in the major groove, where the bases are most accessible.

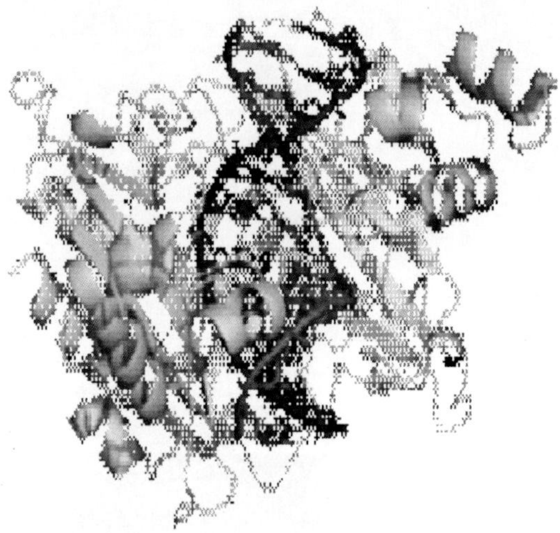

Figure: *The restriction enzyme EcoRV (green) in a complex with its substrate DNA*

DNA-modifying Enzymes

Nucleases and Ligases

Nucleases are enzymes that cut DNA strands by catalyzing the hydrolysis of the phosphodiester bonds. Nucleases that hydrolyse nucleotides from the ends of DNA strands are called exonucleases, while endonucleases cut within strands. The most frequently used nucleases in molecular biology are the restriction endonucleases, which cut DNA at specific sequences.

For instance, the EcoRV enzyme shown to the left recognizes the 6-base sequence 52 -GAT|ATC-32 and makes a cut at the vertical line. In nature, these enzymes protect bacteria against phage infection by digesting the phage DNA when it enters the bacterial cell, acting as part of the restriction modification system.

In technology, these sequence-specific nucleases are used in molecular cloning and DNA fingerprinting.

Enzymes called DNA ligases can rejoin cut or broken DNA strands. Ligases are particularly important in lagging strand DNA replication, as they join together the short segments of DNA produced at the replication fork into a complete copy of the DNA template. They are also used in DNA repair and genetic recombination.

Topoisomerases and Helicases

Topoisomerases are enzymes with both nuclease and ligase activity. These proteins change the amount of supercoiling in DNA. Some of these enzymes work by cutting the DNA helix and allowing one section to rotate, thereby reducing its level of supercoiling; the enzyme then seals the DNA break.

Other types of these enzymes are capable of cutting one DNA helix and then passing a second strand of DNA through this break, before rejoining the helix.

Topoisomerases are required for many processes involving DNA, such as DNA replication and transcription.

Helicases are proteins that are a type of molecular motor. They use the chemical energy in nucleoside triphosphates, predominantly ATP, to break hydrogen bonds between bases and unwind the DNA double helix into single strands. These enzymes are essential for most processes where enzymes need to access the DNA bases.

Polymerases

Polymerases are enzymes that synthesize polynucleotide chains from nucleoside triphosphates. The sequence of their products are copies of existing polynucleotide chains – which are calledtemplates. These enzymes function by adding nucleotides onto the 32 hydroxyl group of the previous nucleotide in a DNA strand. As a consequence, all polymerases work in a 52 to 32 direction. In the active site of these enzymes, the incoming nucleoside triphosphate base-pairs to the template: this allows polymerases to accurately synthesize the complementary strand of their template. Polymerases are classified according to the type of template that they use.

In DNA replication, a DNA-dependent DNA polymerase makes a copy of a DNA sequence. Accuracy is vital in this process, so many of these polymerases have a proofreading activity. Here, the polymerase recognizes the occasional mistakes in the synthesis reaction by the lack of base pairing between the mismatched nucleotides.

If a mismatch is detected, a 32 to 52 exonuclease activity is activated and the incorrect base removed. In most organisms, DNA polymerases function in a large complex called the replisome that contains multiple accessory subunits, such as the DNA clampor helicases.

RNA-dependent DNA polymerases are a specialized class of polymerases that copy the sequence of an RNA strand into DNA. They include reverse transcriptase, which is a viral enzyme involved in the infection of cells by retroviruses, and telomerase, which is required for the replication of telomeres.

Telomerase is an unusual polymerase because it contains its own RNA template as part of its structure.

Transcription is carried out by a DNA-dependent RNA polymerase that copies the sequence of a DNA strand into RNA. To begin transcribing a gene, the RNA polymerase binds to a sequence of DNA called a promoter and separates the DNA strands. It then copies the gene sequence into a messenger RNA transcript until it reaches a region of DNA called the terminator, where it halts and detaches from the DNA.

As with human DNA-dependent DNA polymerases, RNA polymerase II, the enzyme that transcribes most of the genes in the human genome, operates as part of a large protein complexwith multiple regulatory and accessory subunits.

Genetic recombination

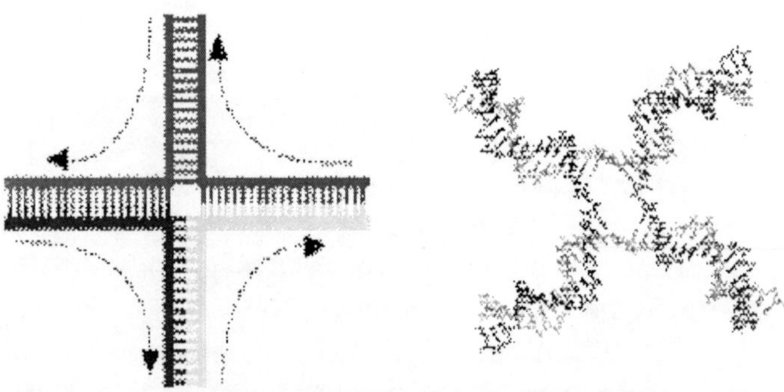

Figure: *Structure of the Holliday junctionintermediate in genetic recombination. The four separate DNA strands are coloured red, blue, green and yellow.*

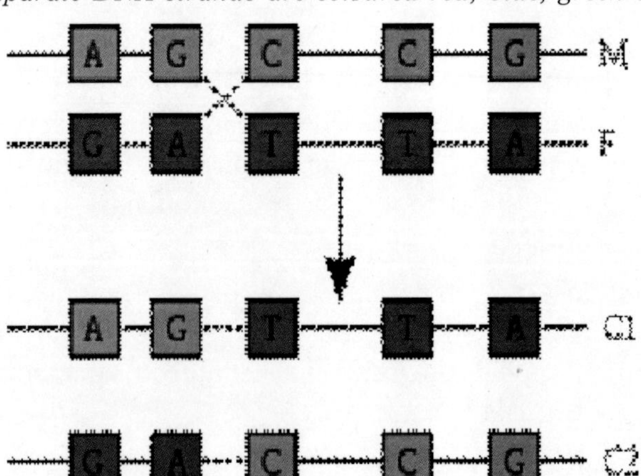

Figure: *Recombination involves the breakage and rejoining of two chromosomes (M and F) to produce two re-arranged chromosomes (C1 and C2).*

A DNA helix usually does not interact with other segments of DNA, and in human cells the different chromosomes even occupy separate areas in the nucleus called "chromosome territories". This physical separation of different chromosomes is important for the ability of DNA to function as a stable repository for information, as one of the few times chromosomes interact is during chromosomal crossover when they recombine. Chromosomal crossover is when two DNA helices break, swap a section and then rejoin.

Recombination allows chromosomes to exchange genetic information and produces new combinations of genes, which increases the efficiency of natural selection and can be important in the rapid evolution of new proteins. Genetic recombination can also be involved in DNA repair, particularly in the cell's response to double-strand breaks.

The most common form of chromosomal crossover is homologous recombination, where the two chroosomes involved share very similar sequences. Nonhomologous recombination can be damaging to cells, as it can produce chromosomal translocations and genetic abnormalities. Te recombination reaction is catalyzed by enzymes known as recobinases, such as RAD51.

The first step in recombination is a duble-stranded break either caused by an endonuclease or damage to the DNA. A series of steps catalyzed in part by the recombnase then leads to joining of the two helices by at least one Holliday junction, in which a segment of a single strand in each helix is annealed to the complementary strand in the ther helix. The Holliday junction is a tetrahedral junction structure that can be moved along the pair of chromosomes, swapping one strand for another. The recombination reaction is then halted by cleavage of the junction and re-ligation of the released DNA.

Evolution

DNA contains the genetic information that allows all modern living things to function, grow and reproduce. However, it is unclear how lng in the 4-billion-year history of life DNA has performed this function, as it has been proposed that the earliest forms of life may have used RNA as their genetic material.

RNA may have acted as the central part of early cell metabolism as it can both transmit genetic information and carry out catalysis as part of ribozymes. This ancient RNA world where nucleic acid would have been used for both catalysis and genetics may have influenced theevolution of the current genetic code based on four nucleotide bases. This would occur, since the number of different bases in such an organism is a trade-off between a small number of bases increasing replication accuracy and a large number of bases increasing the cataltic efficiency of ribozymes.

However, there is no direct evidence of ancient genetic systems, as recovery of DNA from most fossils is impossible. This is because DNA will survive in the environment for less than one million years and slowly degrades into short fragments in solution. Claims for older DNA have been made, most notably a report of the isolation of a viable bacterium from a salt crystal 250 million years old,but these claims are controversial.

On August 8, 2011, a report, based on NASA studies with meteorites found on Earth, was published suggesting building blocks of DNA (adenine, guanine and related organic molecules) may have been formed extraterrestrially in outer space.

Uses in Technology

Genetic Engineering

Methods have been developed to purify DNA from organisms, such as phenol-chloroform extraction, and to manipulate it in the laboratory, such as restriction digests and the polymerase chain reaction. Modern biology and biochemistry make intensive use of these techniques in recombinant DNA technology. Recombinant DNA is a man-made DNA sequence that has been assembled from other DNA sequences.

They can be transformed into organisms in the form of plasmids or in the appropriate format, by using a viral vector. The genetically modified organisms produced can be used to produce products such as recombinant proteins, used in medical research, or be grown in agriculture.

Forensics

Forensic scientists can use DNA in blood, semen, skin, saliva or hair found at a crime scene to identify a matching DNA of an individual, such as a perpetrator. This process is formally termed DNA profiling, but may also be called "genetic fingerprinting". In DNA profiling, the lengths of variable sections of repetitive DNA, such as short tandem repeats and minisatellites, are compared between people. This method is usually an extremely reliable technique for identifying a matching DNA. However, identification can be complicated if the scene is contaminated with DNA from several people. DNA profiling was developed in 1984 by British geneticist Sir Alec Jeffreys, and first used in forensic science to convict Colin Pitchfork in the 1988 Enderby murders case.

People convicted of certain types of crimes may be required to provide a sample of DNA for a database. This has helped investigators solve old cases where only a DNA sample was obtained from the scene. DNA profiling can also be used to identify victims of mass casualty incidents. On the other hand, many convicted people have been released from prison on the basis of DNA techniques, which were not available when a crime had originally been committed.

Bioinformatics

Bioinformatics involves the manipulation, searching, and data mining of biological data, and this includes DNA sequence data. The development of techniques to store and search DNA sequences have led to widely applied advances in computer science, especially string searching algorithms, machine learning and database theory. String searching or matching algorithms, which find an occurrence of a sequence of letters inside a larger sequence of letters, were developed to search for specific sequences of nucleotides.

The DNA sequenced may be aligned with other DNA sequences to identify homologous sequences and locate the specific mutations that make them distinct.

These techniques, especially multiple sequence alignment, are used in studying phylogeneticrelationships and protein function. Data sets representing entire genomes' worth of DNA sequences, such as those produced by the Human Genome Project, are difficult to use without the annotations that identify the locations of genes and regulatory elements on each chromosome.

Regions of DNA sequence that have the characteristic patterns associated with protein- or RNA-coding genes can be identified by gene finding algorithms, which allow researchers to predict the presence of particular gene products and their possible functions in an organism even before they have been isolated experimentally. Entire genomes may also be compared which can shed light on the evolutionary history of particular organism and permit the examination of complex evolutionary events.

DNA nanotechnology

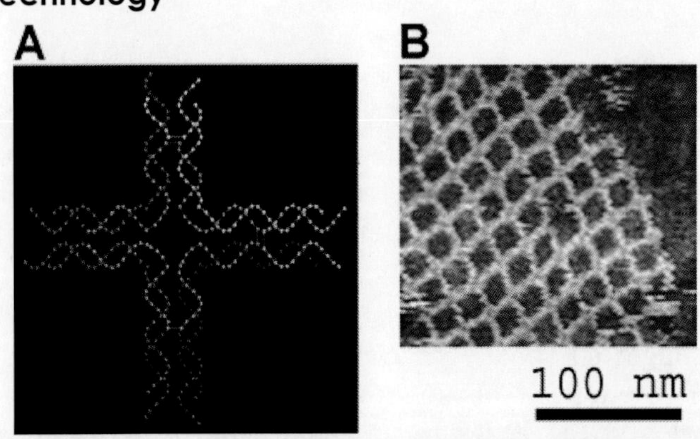

The DNA structure at left (schematic shown) will self-assemble into the structure visualized by atomic force microscopy at right. DNA nanotechnology is the field that seeks to design nanoscale structures using the molecular recognitionproperties of DNA molecules. Image from Strong, 2004.

DNA Nanotechnology

DNA nanotechnology uses the unique molecular recognition properties of DNA and other nucleic acids to create self-assembling branched DNA complexes with useful properties. DN is thus used as a structural material rather than as a carrier of biological information.

This has led to the creation of two-dimensional periodic lattices (both tile-based as well as using the "DNA origami" method) as well as three-dimensional structures in the shapes of polyhedra. Nanomechanical devices and algorithmic self-assembly have also been demonstrated, and these DNA structures have been used to template the arrangement of other molecules such as gold nanoparticles and streptavidin proteins.

History and Anthropology

Because DNA collects mutations over time, which are then inherited, it contains historical information, and, by comparing DNA sequences, geneticists can infer the evolutionary history of organisms, their phylogeny. This field of phylogenetics is a powerful tool in evolutionary biology. If DNA sequences within a species are compared, population geneticists can learn the history of particular populations. This can be used in studies ranging from ecological genetics to anthropology; For example, DNA evidence is being used to try to identify the Ten Lost Tribes of Israel.

DNA has also been used to look at modern family relationships, such as establishing family relationships between the descendants of Sally Hemings and Thomas Jefferson. This usage is closely related to the use of DNA in criminal investigations detailed above. Indeed, some criminal investigations have been solved when DNA from crime scenes has matched relatives of the guilty individual.

History of DNA Research

DNA was first isolated by the Swiss physician Friedrich Miescher who, in 1869, discovered a microscopic substance in the pus of discarded surgical bandages. As it resided in the nuclei of cells, he called it "nuclein". In 1878, Albrecht Kossel isolated the non-protein component of "nuclein", nucleic acid, and later isolated its five primary nucleobases.

In 1919, Phoebus Levene identified the base, sugar and phosphate nucleotide unit. Levene suggested that DNA consisted of a string of nucleotide units linked together through the phosphate groups. However, Levene thought the chain was short and the bases repeated in a fixed order. In 1937 William Astbury produced the first X-ray diffraction patterns that showed that DNA had a regular structure.

Rosalind Franklin used X-ray crystallography to help visualize the structure of DNA.

In 1927 Nikolai Koltsov proposed that inherited traits would be inherited via a "giant hereditary molecule" which would be made up of "two mirror strands that would replicate in a semi-conservative fashion using each strand as a template". In 1928, Frederick Griffith discovered that traits of the "smooth" form of the Pneumococcus could be transferred to the "rough" form of the same bacteria by mixing killed "smooth" bacteria with the live "rough" form. This system provided the first clear suggestion that DNA carries genetic information—the Avery–MacLeod–McCarty experiment—when Oswald Avery, along with coworkers Colin MacLeod and Maclyn McCarty, identified DNA as the transforming principle in 1943. DNA's role in heredity was confirmed in 1952, when Alfred Hershey and Martha Chase in the Hershey–Chase experiment showed that DNA is the genetic material of the T2 phage.

In 1953, James D. Watson and Francis Crick suggested what is now accepted as the first correct double-helix model of DNA structure in the journal Nature.

Their double-helix, molecular model of DNA was then based on a single X-ray diffraction image (labeled as "Photo 51") taken by Rosalind Franklin and Raymond Gosling in May 1952, as well as the information that the DNA bases are paired — also obtained through private communications from Erwin Chargaff in the previous years. Chargaff's rules played a very important role in establishing double-helix configurations for B-DNA as well as A-DNA.

Experimental evidence supporting the Watson and Crick model were published in a series of five articles in the same issue of Nature. Of these, Franklin and Gosling's paper was the first publication of their own X-ray diffraction data and original analysis method that partially supported the Watson and Crick model;this issue also contained an article on DNA structure by Maurice Wilkins and two of his colleagues, whose analysis and in vivo B-DNA X-ray patterns also supported the presence in vivo of the double-helical DNA configurations as proposed by Crick and Watson for their double-helix molecular model of DNA in the previous two pages of Nature. In 1962, after Franklin's death, Watson, Crick, and Wilkins jointly received the Nobel Prize in Physiology or Medicinc. However, Nobel rules of the time allowed only living recipients, but a vigorous debate continues on who should receive credit for the discovery.

In an influential presentation in 1957, Crick laid out the central dogma of molecular biology, which foretold the relationship between DNA, RNA, and proteins, and articulated the "adaptor hypothesis". Final confirmation of the replication mechanism that was implied by the double-helical structure followed in 1958 through the Meselson–Stahl experiment. Further work by Crick and coworkers showed that the genetic code was based on non-overlapping triplets of bases, called codons, allowing Har Gobind Khorana, Robert W. Holley and

Marshall Warren Nirenberg to decipher the genetic code. These findings represent the birth of molecular biology.

RNA

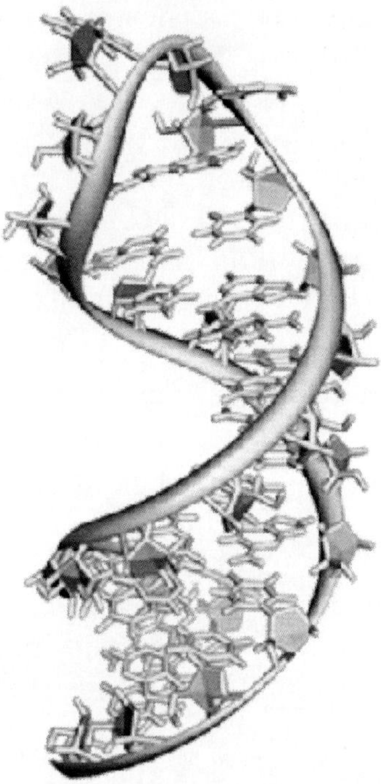

Figure: *A hairpin loop from a pre-mRNA. Highlighted are the nucleobases (green) and the ribose-phosphate backbone (blue).*

Ribonucleic acid (English pronunciation: /rajbu.njuÐÌklej.hk Èæsjd/), or RNA, is one of the three major macromolecules (along with DNA and proteins) that are essential for all known forms of life. Like DNA, RNA is made up of a long chain of components called nucleotides.

Each nucleotide consists of a nucleobase (sometimes called a nitrogenous base), a ribose sugar, and a phosphate group. The sequence of nucleotides allows RNA to encode genetic information. All cellular organisms use messenger RNA(mRNA) to carry the genetic information that directs the synthesis of proteins. In addition, some viruses use RNA instead of DNA as their genetic material; perhaps a reflection of the suggested key role of RNA in the evolutionary history of life on Earth.

Like proteins, some RNA molecules play an active role in cells by catalyzing biological reactions, controlling gene expression, or sensing and communicating responses to cellular signals.

One of these active processes is protein synthesis, a universal function whereby mRNA molecules direct the assembly of proteins on ribosomes.

This process uses transfer RNA (tRNA) molecules to deliver amino acids to the ribosome, where ribosomal RNA (rRNA) links amino acids together to form proteins. In 2011, it was proved that the methylation of mRNA has a critical role in human energy homeostasis. This opens up the field of RNA epigenetics.

The chemical structure of RNA is very similar to that of DNA, with two differences: (a) RNA contains the sugar ribose, while DNA contains the slightly different sugar deoxyribose (a type of ribose that lacks one oxygen atom), and (b) RNA has the nucleobase uracil while DNA contains thymine. Uracil and thymine have similar base-pairing properties.

Unlike DNA, most RNA molecules are single-stranded. Single-stranded RNA molecules adopt very complex three-dimensional structures, since they are not restricted to the repetitive double-helical form of double-stranded DNA.

RNA is made within living cells by RNA polymerases, enzymes that act to copy a DNA or RNA template into a new RNA strand through processes known as transcription or RNA replication, respectively.

Comparison with DNA

Figure: *Three-dimensional representation of the50S ribosomal subunit. RNA is in ochre, protein in blue. The active site is in the middle (red).*

RNA and DNA are both nucleic acids, but differ in three main ways. First, unlike double-stranded DNA, RNA is a single-stranded molecule in many of its biological roles and has a much shorter chain of nucleotides. Second, while DNA contains deoxyribose, RNA contains ribose (in deoxyribose there is no hydroxyl group attached to the pentose ring in the 2' position).

These hydroxyl groups make RNA less stable than DNA because it is more prone tohydrolysis. Third, the complementary base to adenine is not thymine, as it is in DNA, but rather uracil, which is an unmethylated form of thymine.

Like DNA, most biologically active RNAs, including mRNA, tRNA, rRNA, snRNAs, and other non-coding RNAs, contain self-complementary sequences that allow parts of the RNA to fold and pair with itself to form double helices. Analysis of these RNAs has revealed that they are highly structured.

Unlike DNA, their structures do not consist of long double helices but rather collections of short helices packed together into structures akin to proteins.

In this fashion, RNAs can achieve chemical catalysis, like enzymes. For instance, determination of the structure of the ribosome—an enzyme that catalyzes peptide bond formation—revealed that its active site is composed entirely of RNA.

Structure

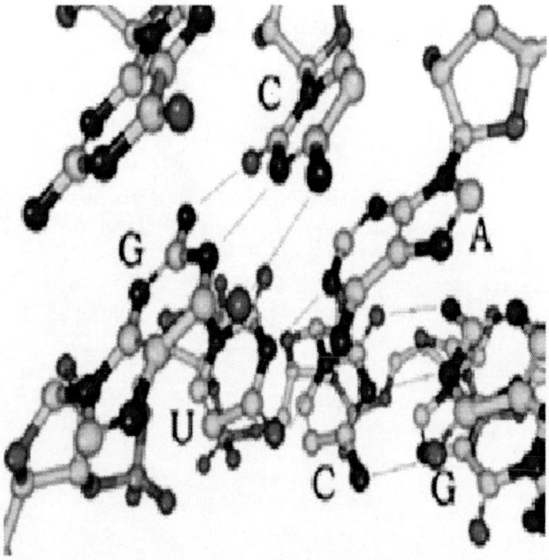

Figure: *Watson-Crick base pairs in a siRNA(hydrogen atoms are not shown)*

Each nucleotide in RNA contains a ribose sugar, with carbons numbered 1' through 5'. A base is attached to the 1' position, in general, adenine (A), cytosine(C), guanine (G), or uracil (U). Adenine and guanine are purines, cytosine, and uracil are pyrimidines. A phosphate group is attached to the 3' position of one ribose and the 5' position of the next. The phosphate groups have a negative charge each at physiological pH, making RNA a charged molecule (polyanion).

The bases may form hydrogen bonds between cytosine and guanine, between adenine and uracil and between guanine and uracil.

However, other interactions are possible, such as a group of adenine bases binding to each other in a bulge, or the GNRA tetraloop that has a guanine–adenine base-pair.

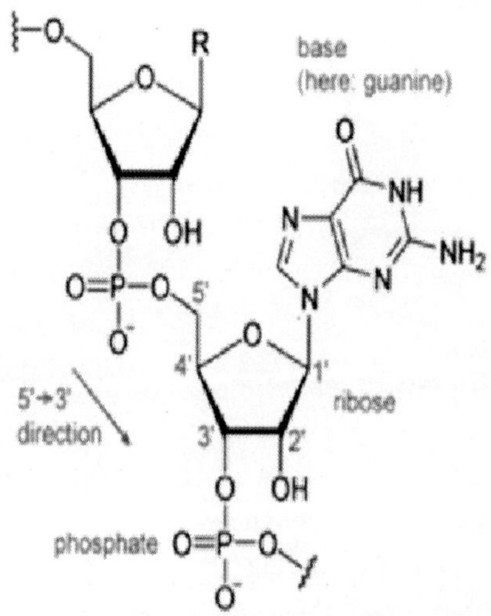

***Fig. 17:** Chemical structure of RNA*

An important structural feature of RNA that distinguishes it from DNA is the presence of a hydroxyl group at the 2' position of the ribose sugar. The presence of this functional group causes the helix to adopt the A-form geometry rather than the B-form most commonly observed in DNA.

This results in a very deep and narrow major groove and a shallow and wide minor groove. A second consequence of the presence of the 2'-hydroxyl group is that in conformationally flexible regions of an RNA molecule (that is, not involved in formation of a double helix), it can chemically attack the adjacent phosphodiester bond to cleave the backbone. RNA is transcribed with only four bases (adenine, cytosine, guanine and uracil), but these bases and attached sugars can be modified in numerous ways as the RNAs mature.

Pseudouridine (Ø), in which the linkage between uracil and ribose is changed from a C–N bond to a C–C bond, and ribothymidine (T) are found in various places (the most notable ones being in the TØC loop of tRNA).

Another notable modified base is hypoxanthine, a deaminated adenine base whose nucleoside is called inosine (I). Inosine plays a key role in the wobble hypothesisof the genetic code.

There are nearly 100 other naturally occurring modified nucleosides, of which pseudouridine and nucleosides with 2'-O-methylribose are the most common.

The specific roles of many of these modifications in RNA are not fully understood. However, it is notable that, in ribosomal RNA, many of the post-transcriptional modifications occur in highly functional regions, such as the peptidyl transferase center and the subunit interface, implying that they are important for normal function.

The functional form of single stranded RNA molecules, just like proteins, frequently requires a specific tertiary structure. The scaffold for this structure is provided by secondary structural elements that are hydrogen bonds within the molecule.

This leads to several recognizable "domains" of secondary structure like hairpin loops, bulges, and internal loops. Since RNA is charged, metal ions such as Mg are needed to stabilise many secondary and tertiary structures.

Synthesis

Synthesis of RNA is usually catalyzed by an enzyme—RNA polymerase—using DNA as a template, a process known as transcription. Initiation of transcription begins with the binding of the enzyme to a promoter sequence in the DNA (usually found "upstream" of a gene).

The DNA double helix is unwound by the helicase activity of the enzyme.

The enzyme then progresses along the template strand in the 3' to 5' direction, synthesizing a complementary RNA molecule with elongation occurring in the 5' to 3' direction.

The DNA sequence also dictates where termination of RNA synthesis will occur.

RNAs are often modified by enzymes after transcription. For example, a poly(A) tail and a 5' cap are added to eukaryotic pre-mRNA and introns are removed by the spliceosome.

There are also a number of RNA-dependent RNA polymerases that use RNA as their template for synthesis of a new strand of RNA. For instance, a number of RNA viruses (such as poliovirus) use this type of enzyme to replicate their genetic material.

Also, RNA-dependent RNA polymerase is part of the RNA interference pathway in many organisms.

Types of RNA

Messenger RNA (mRNA) is the RNA that carries information from DNA to the ribosome, the sites of protein synthesis (translation) in the cell. The coding sequence of the mRNA determines the amino acid sequence in the protein that is produced. Many RNAs do not code for protein however (about 97% of the transcriptional output is non-protein-coding in eukaryotes).

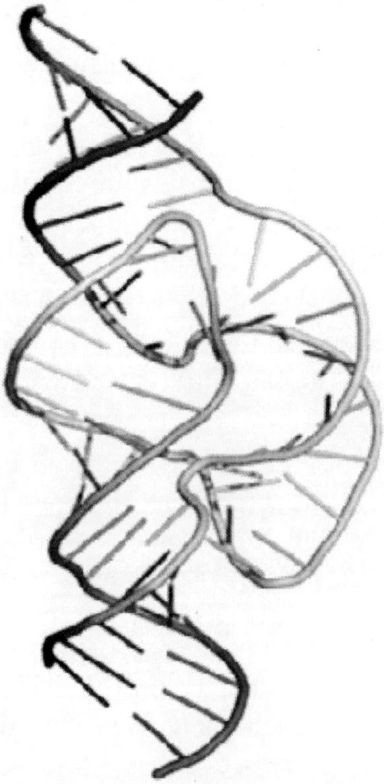

Figure: *Structure of a hammerhead ribozyme, a ribozyme that cuts RNA*

These so-called non-coding RNAs ("ncRNA") can be encoded by their own genes (RNA genes), but can also derive from mRNA introns. The most prominent examples of non-coding RNAs are transfer RNA (tRNA) and ribosomal RNA (rRNA), both of which are involved in the process of translation.

There are also non-coding RNAs involved in gene regulation, RNA processing and other roles. Certain RNAs are able to catalyse chemical reactions such as cutting and ligating other RNA molecules, and the catalysis of peptide bond formation in the ribosome; these are known as ribozymes.

In translation

Messenger RNA (mRNA) carries information about a protein sequence to the ribosomes, the protein synthesis factories in the cell. It is coded so that every three nucleotides (a codon) correspond to one amino acid.

In eukaryotic cells, once precursor mRNA (pre-mRNA) has been transcribed from DNA, it is processed to mature mRNA. This removes its introns—non-coding sections of the pre-mRNA. The mRNA is then exported from the nucleus to the cytoplasm, where it is bound to ribosomes and translated into its corresponding protein form with the help of tRNA. In prokaryotic cells, which do not have nucleus and cytoplasm compartments, mRNA can bind to ribosomes while it is being transcribed from DNA. After a certain amount of time the message degrades into its component nucleotides with the assistance of ribonucleases.

Transfer RNA (tRNA) is a small RNA chain of about 80 nucleotides that transfers a specific amino acid to a growing polypeptide chain at the ribosomal site of protein synthesis during translation. It has sites for amino acid attachment and an anticodon region for codon recognition that binds to a specific sequence on the messenger RNA chain through hydrogen bonding.

Ribosomal RNA (rRNA) is the catalytic component of the ribosomes. Eukaryotic ribosomes contain four different rRNA molecules: 18S, 5.8S, 28S and 5S rRNA.

Three of the rRNA molecules are synthesized in the nucleolus, and one is synthesized elsewhere. In the cytoplasm, ribosomal RNA and protein combine to form a nucleoprotein called a ribosome. The ribosome binds mRNA and carries out protein synthesis. Several ribosomes may be attached to a single mRNA at any time. Nearly all the RNA found in a typical eukaryotic cell is rRNA. Transfer-messenger RNA (tmRNA) is found in many bacteria and plastids. It tags proteins encoded by mRNAs that lack stop codons for degradation and prevents the ribosome from stalling.

Regulatory RNAs

Several types of RNA can downregulate gene expression by being complementary to a part of an mRNA or a gene's DNA. MicroRNAs (miRNA; 21-22 nt) are found in eukaryotes and act through RNA interference (RNAi), where an effector complex of miRNA and enzymes can cleave complementary mRNA, block the mRNA from being translated, or accelerate its degradation.

While small interfering RNAs (siRNA; 20-25 nt) are often produced by breakdown of viral RNA, there are also endogenous sources of siRNAs.

siRNAs act through RNA interference in a fashion similar to miRNAs. Some miRNAs and siRNAs can cause genes they target to be methylated, thereby decreasing or increasing transcription of those genes. Animals have Piwi-interacting RNAs (piRNA; 29-30 nt) which are active in germline cells and are thought to be a defense against transposons and play a role in gametogenesis.

Many prokaryotes have CRISPR RNAs, a regulatory system similar to RNA interference.

Antisense RNAs are widespread; most downregulate a gene, but a few are activators of transcription.One way antisense RNA can act is by binding to an mRNA, forming double-stranded RNA that is enzymatically degraded. There are many long noncoding RNAs that regulate genes in eukaryotes, one such RNA is Xist, which coats one X chromosome in female mammals and inactivates it.

An mRNA may contain regulatory elements itself, such as riboswitches, in the 5' untranslated region or 3' untranslated region; these cis-regulatory elements regulate the activity of that mRNA. The untranslated regions can also contain elements that regulate other genes.

In RNA Processing

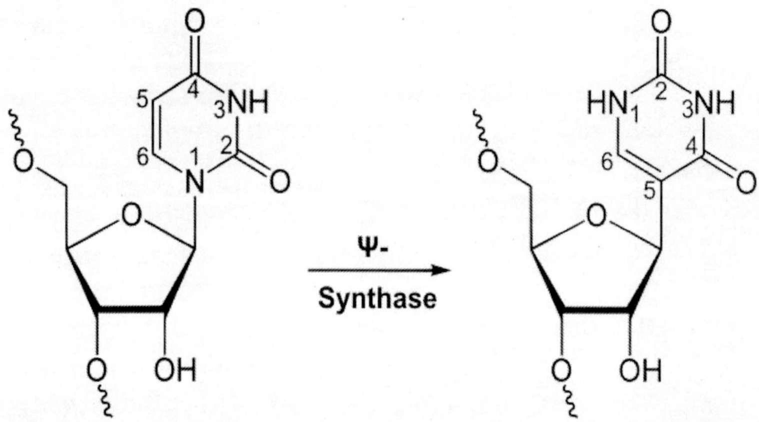

Figure: *Uridine to pseudouridine is a common RNA modification.*

Many RNAs are involved in modifying other RNAs. Introns are spliced out of pre-mRNA by spliceosomes, which contain several small nuclear RNAs(snRNA), or the introns can be ribozymes that are spliced by themselves.

RNA can also be altered by having its nucleotides modified to other nucleotides than A, C, G and U.

In eukaryotes, modifications of RNA nucleotides are generally directed by small nucleolar RNAs (snoRNA; 60-300 nt),found in the nucleolus and cajal bodies. snoRNAs associate with enzymes and guide them to a spot on an RNA by basepairing to that RNA.

These enzymes then perform the nucleotide modiication. rRNAs and tRNAs are extensively modified, but snRNAs and mRNAs can also be the target of base modification. RNA can also be methylated.

RNA Genomes

Like DNA, RNA can carry genetic information. RNA viruses have genomes composed of RNA which encodes a number of proteins.

The viral genome is replicated by some of those proteins, while other proteins protect the genome as the virus particle moves to a new host cell. Viroids are another group of pathogens, but they consist only of RNA, do not encode any protein and are replicated by a host plant cell's polymerase.

In Reverse Transcription

Reverse transcribing viruses replicate their genomes by reverse transcribing DNA copies from their RNA; these DNA copies are then transcribed to new RNA. Retrotransposons also spread by copying DNA and RNA from one another, and telomerase contains an RNA that is used as template for building the ends of eukaryotic chromosomes.

Double-stranded RNA

Double-stranded RNA (dsRNA) is RNA with two complementary strands, similar to the DNA found in all cells. dsRNA forms the genetic material of some viruses (double-stranded RNA viruses). Double-stranded RNA such as viral RNA or siRNA can trigger RNA interference in eukaryotes, as well as interferon response in vertebrates.

Key Discoveries in RNA Biology

Research on RNA has led to many important biological discoveries and numerous Nobel Prizes. Nucleic acids were discovered in 1868 by Friedrich Miescher, who called the material 'nuclein' since it was found in the nucleus.

It was later discovered that prokaryotic cells, which do not have a nucleus, also contain nucleic acids. The role of RNA in protein synthesis was suspected already in 1939. Severo Ochoa won the 1959 Nobel Prize in Medicine (shared with Arthur Kornberg) after he discovered an enzyme that can synthesize RNA in the laboratory. Ironically, the enzyme discovered by Ochoa (polynucleotide phosphorylase) was laer shown to be responsible for RNA degradation, not RNA synthesis.

The sequence of the 77 nucleotides of a yeast tRNA was found by Robert W. Holley in 1965, winning Holley the 1968 Nobel Prize in Medicine (shared with Har Gobind Khorana and

Marshall Nirenberg). In 1967, Carl Woese hypothesized that RNA might be catalytic and suggested that the earliest forms of life (self-replicating molecules) could have relied on RNA both to carry genetic information and to catalyze biochemical reactions—an RNA world.

During the early 1970s, retroviruses and reverse transcriptase were discovered, showing for the first time that enzymes could copy RNA into DNA (the opposite of the usual route for transmission of genetic information).

For this work, David Baltimore, Renato Dulbecco and Howard Temin were awarded a Nobel Prize in 1975. In 1976, Walter Fiers and his team determined the first complete nucleotide sequence of an RNA virus genome, that of bacteriophage MS2.

In 1977, introns and RNA splicing were discovered in both mammalian viruses and in cellular genes, resulting in a 1993 Nobel to Philip Sharp and Richard Roberts.

Catalytic RNA molecules (ribozymes) were discovered in the early 1980s, leading to a 1989 Nobel award to Thomas Cech and Sidney Altman. In 1990 it was found in petunia that introduced genes can silence similar genes of the plant's own, now known to be a result of RNA interference.

At about the same time, 22 nt long RNAs, now called microRNAs, were found to have a role in the development of C. elegans. Studies on RNA interference gleaned a Nobel Prize for Andrew Fireand Craig Mello in 2006, and another Nobel was awarded for studies on transcription of RNA to Roger Kornberg in the same year. The discovery of gene regulatory RNAs has led to attempts to develop drugs made of RNA, such as siRNA, to silence genes.

Nucleic Acid

Nucleic acids are biological molecules essential for life, and include DNA (deoxyribonucleic acid) and RNA (ribonucleic acid). Together with proteins, nucleic acids make up the most important macro molecules; each is found in abundance in all living things, where they function in encoding, transmitting and expressing genetic information. Nucleic acids were first discovered by Friedrich Miescher in 1871.

Experimental studies of nucleic acids constitute a major part of modern biological and medical research, and form a foundation for genome and forensic science, as well as thebiotechnology and pharmaceutical industries.

Occurrence and Nomenclature

The term nucleic acid is the overall name for DNA and RNA, members of a family of biopolymers, and is synonymous with polynucleotide. Nucleic acids were named for their initial discovery within the cell nucleus, and for the presence of phosphate groups (related to phosphoric acid).

Although first discovered within the nucleus of eukaryotic cells, nucleic acids are now known to be found in all life forms, including within bacteria, archaea, mitochondria, chloroplasts, viruses and viroids.

All living cells and organelles contain both DNA and RNA, while viruses contain either DNA or RNA, but usually not both. The basic component of biological nucleic acids is the nucleotide, each of which contains a pentose sugar (ribose or deoxyribose), a phosphate group, and a nucleobase. Nucleic acids are also generated within the laboratory, through the use of enzymes (DNA and RNA polymerases) and by solid-phase chemical synthesis. The chemical methods also enable the generation of altered nucleic acids that are not found in nature, for example peptide nucleic acids.

Molecular composition and size

Nucleic acids can vary in size, but are generally very large molecules. Indeed, DNA molecules are probably the largest individual molecules known. Well-studied biological nucleic acid molecules range in size from 21 nucleotides (small interfering RNA) to large chromosomes (human chromosome 1 is a single molecule that contains 247 million base pairs).

In most cases, naturally occurring DNA molecules are double-stranded and RNA molecules are single-stranded. There are numerous exceptions, however—some viruses have genomes made of double-stranded RNA and other viruses have single-stranded DNA genomes, and, in some circumstances, nucleic acid structures with three or four strands can form. Nucleic acids are linear polymers (chains) of nucleotides. Each nucleotide consists of three components: a purine or pyrimidine nucleobase (sometimes termed nitrogenous base or simply base), apentose sugar, and a phosphate group.

The substructure consisting of a nucleobase plus sugar is termed a nucleoside. Nucleic acid types differ in the structure of the sugar in their nucleotides - DNA contains 2'-deoxyribose while RNA contains ribose (where the only difference is the presence of a hydroxyl group).

Also, the nucleobases found in the two nucleic acid types are different: aenine,cytosine, andguanine are found in both RNA and DNA, while thymine occurs in DNA and uracil occurs in RNA.

The sugars and phosphates in nucleic acids are connected to each other in an alternating chain (sugar-phosphate backbone) through phosphodiester linkages. In conventional nomenclature, the carbons to which the phosphate groups attach are the 3'-end and the 5'-end carbons of the sugar.

This gives nucleic acids directionality, and the ends of nucleic acid molecules are referred to as 5'-end and 3'-end. The nucleobases are joined to the sugars via an N-glycosidic linkage involving a nucleobase ring nitrogen (N-1 for pyrimidines and N-9 for purines) and the 1' carbon of the pentose sugar ring.

Non-standard nucleosides are also found in both RNA and DNA and usually arise from modification of the standard nucleosides within the DNA molecule or the primary (initial) RNA transcript.

Transfer RNA (tRNA) molecules contain a particularly large number of modified nucleosides.

Topology

Double-stranded nucleic acids are made up of complementary sequences, in which extensive Watson-Crick base pairing results in the a highly repeated and quite uniform double-helical three-dimensional structure. In contrast, single-stranded RNA and DNA molecules are not constrained to a regular double helix, and can adopt highly complex three-dimensional structures that are based on short stretches of intramolecular base-paired sequences that include both Watson-Crick and noncanonical base pairs, as well as a wide range of complex tertiary interactions.Nucleic acid molecules are usually unbranched, and may occur as linear and circular molecules.

For example, bacterial chromosomes, plasmids, mitochondrial DNA and chloroplast DNA are usually circular double-stranded DNA molecules, while chromosomes of the eukaryotic nucleus are usually linear double-stranded DNA molecules.

Most RNA molecules are linear, single-stranded molecules, but both circular and branched molecules can result from RNA splicing reactions.

Nucleic Acid Sequences

One DNA or RNA molecule differs from another primarily in the sequence of nucleotides.

Nucleotide sequences are of great importance in biology, since they carry the ultimate instructions that encode all biological molecules, molecular assemblies, subcellular and cellular structures, organs and organisms, and directly enable cognition, memory and behavior.

Enormous efforts have gone into the development of experimental methods to determine the nucleotide sequence of biological DNA and RNA molecules, and today hundreds of millions of nucleotides are sequenceddaily at genome centers and smaller laboratories worldwide.

Types of Nucleic Acis

Deoxyribonucleic Acid: Deoxyribonucleic acid is a nucleic acid that contains the genetic instructions used in the deveopment and functioning of all known living organisms.

The main role of DNA molecules is the long-term storage of information and DNA is often compared to a set of blueprints, since it contains the instructions needed to construct other components of cells, such as proteins and RNA molecules.

The DNA segments that carry this genetic information are called genes, but other DNA sequences have structural purposes, or are involved in regulating the use of this genetic information.hi

Ribonucleic Acid

Ribonucleic acid (RNA) functions in converting genetic information from genes into the amino acid sequences of proteins. The three universal types of RNA include transfer RNA (tRNA), messenger RNA (mRNA), and ribosomal

RNA (rRNA). Messenger RNA acts to carry genetic sequence information between DNA and ribosomes, directing protein synthesis.

Ribosomal RNA is a major component of the ribosome, and catalyzes peptide bond formation. Transfer RNA serves as the carrier molecule for amino acids to be used in protein synthesis, and is responsible for decoding the mRNA. In addition, many other classes of RNA are now known.

Artificial Nucleic Acid Analogs

Artificial nucleic acid analogs have been designed and synthesized by chemists, and include peptide nucleic acid, morpholino- and locked nucleic acid, as well as glycol nucleic acid and threose nucleic acid. Each of these is distinguished from naturally-occurring DNA or RNA by changes to the backbone of the molecule.

Nucleic Acid Simulation Software

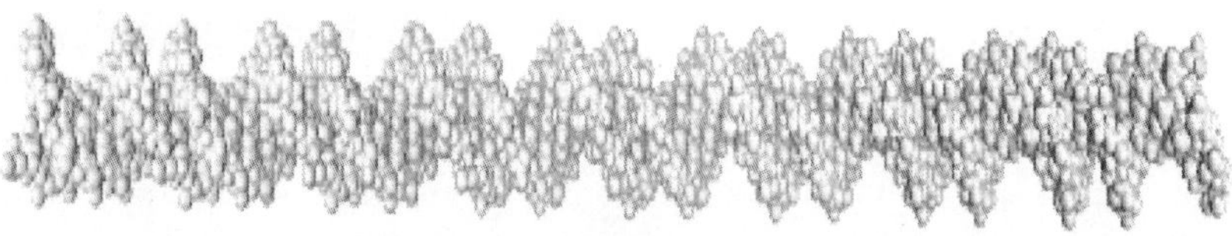

Figure: *This is a list of computer programs that are used for nucleic acids simulations.*

Min - Optimization, MD - Molecular dynamics, MC - Monte Carlo, REM - Replica exchange method,

Crt - Cartesian coordinates. Int - Internal coordinates Exp - Explicit water. Imp - Implicit water.

Lig - Ligands interactions. HA - Hardware accelerated.

Nucleic Acid Methods

Nucleic acid methods are the techniques used to study nucleic acids (DNA and RNA).

Purification

- Phenol-chloroform extraction
- minicolumn purification
- RNA extraction

Quantification

- Abundance in weight: spectroscopic quantification
- Absolute abundance in number: Q-PCR
- high-throughput relative abundance DNA microarray
- high-throughput absolute abundance SAGE
- Size: Gel electrophoresis

Synthesis

- De novo: Oligonucleotide synthesis
- Amplification: PCR

Kinetics

- Dual polarization interferometry
- QCM-D

Other

- Nucleic acid simulations
- DNA sequencing
- Bisulfite sequencing
- Expression cloning
- Southern blot
- Northern blot
- Sucrose gradient centrifugation
- Radioactivity in biological research
- Lab-on-a-chip
- Nuclear run-on assay
- Fluorescent in situ hybridization
- several Bioinformatics methods, such as RNA structure prediction

8

DNA Sequencing

DNA sequencing includes several methods and technologies that are used for determining the order of the nucleotide bases—adenine, guanine, cytosine, and thymine—in a molecule of DNA.

Knowledge of DNA sequences has become indispensable for basic biological research, other research branches utilizing DNA sequencing, and in numerous applied fields such as diagnostic, biotechnology, forensic biology and biological systematics.

The advent of DNA sequencing has significantly accelerated biological research and discovery. The rapid speed of sequencing attained with modern DNA sequencing technology has been instrumental in the sequencing of the human genome, in the Human Genome Project.

Related projects, often by scientific collaboration across continents, have generated the complete DNA sequences of many animal, plant, and microbial genomes.

The first DNA sequences were obtained in the early 1970s by academic researchers using laborious methods based on two-dimensional chromatography.

Following the development of dye-based sequencing methods with automated analysis, DNA sequencing has become easier and orders of magnitude faster.

RNA sequencing was one of the earliest forms of nucleotide sequencing. The major landmark of RNA sequencing is the sequence of the first complete gene and the complete genome of Bacteriophage MS2, identified and published by Walter Fiers and his coworkers at the University of Ghent (Ghent, Belgium), between 1972 and 1976.

Prior to the development of rapid DNA sequencing methods in the early 1970s by Frederick Sanger at the University of Cambridge, in England and Walter Gilbert and Allan Maxam at Harvard, a number of laborious methods were used. For instance, in 1973, Gilbert and Maxam reported the sequence of 24 basepairs using a method known as wandering-spot analysis.

The chain-termination method developed by Sanger and coworkers in 1977 soon became the method of choice, owing to its relative ease and reliability. It involves separating DNA bases from different DNA fragments.

Maxam–Gilbert Sequencing

In 1976–1977, Allan Maxam and Walter Gilbert developed a DNA sequencing method based on chemical modification of DNA and subsequent cleavage at specific bases. Although Maxam and Gilbert published their chemical sequencing method two years after the ground-breaking paper of Sanger and Coulson on plus-minus sequencing, Maxam–Gilbert sequencing rapidly became more popular, since purified DNA could be used directly, while the initial Sanger method required that each read start be cloned for production of single-stranded DNA.

However, with the improvement of the chain-termination method (see below), Maxam-Gilbert sequencing has fallen out of favour due to its technical complexity prohibiting its use in standard molecular biology kits, extensive use of hazardous chemicals, and difficulties with scale-up.

The method requires radioactive labeling at one 5' end of the DNA (typically by a kinase reaction using gamma-P ATP) and purification of the DNA fragment to be sequenced. Chemical treatment generates breaks at a small proportion of one or two of the four nucleotide bases in each of four reactions (G, A+G, C, C+T).

For example, the purines (A+G) are depurinated using formic acid, the guanines (and to some extent the adenines) are methylated by dimethyl sulfate, and the pyrimidines (C+T) are methylated using hydrazine.

The addition of salt (sodium chloride) to the hydrazine reaction inhibits the methylation of thymine for the C-only reaction. The modified DNAs are then cleaved by hot piperidine at the position of the modified base. The concentration of the modifying chemicals is controlled to introduce on average one modification per DNA molecule. Thus a series of labeled fragments is generated, from the radiolabeled end to the first "cut" site in each molecule.

The fragments in the four reactions are electrophoresed side by side in denaturing acrylamide gels for size separation. To visualize the fragments, the gel is exposed to X-ray film for autoradiography, yielding a series of dark bands each corresponding to a radiolabeled DNA fragment, from which the sequence may be inferred.

Also sometimes known as "chemical sequencing", this method led to the Methylation Interference Assay used to map DNA-binding sites for DNA-binding proteins.

Chain-termination Methods

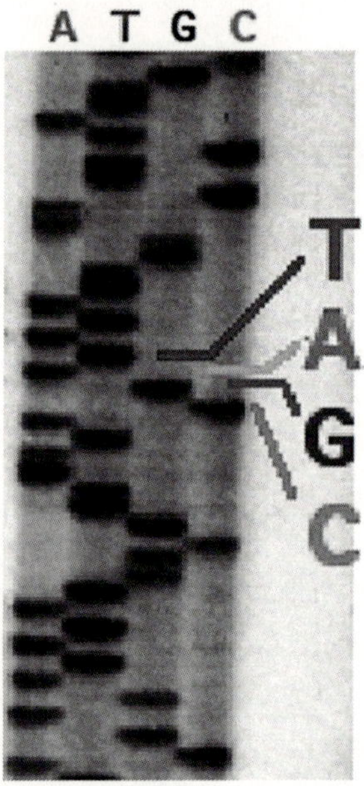

Figure: *Part of a radioactively labelled sequencing gel*

Because the chain-terminator method (or Sanger method after its developer Frederick Sanger) is more efficient and uses fewer toxic chemicals and lower amounts of radioactivity than the method of Maxam and Gilbert, it rapidly became the method of choice. The key principle of the Sanger method was the use of dideoxy nucleotide triphosphates (ddNTPs) as DNA chain terminators.

The classical chain-termination method requires a single-stranded DNA template, a DNA primer, a DNA polymerase, normal deoxynucleotidetriphosphates (dNTPs), and modified nucleotides (dideoxyNTPs) that terminate DNA strand elongation. These ddNTPs will also be radioactively or fluorescently labelled for detection in automated sequencing machines.

The DNA sample is divided into four separate sequencing reactions, containing all four of the standard deoxynucleotides (dATP, dGTP, dCTP and dTTP) and the DNA polymerase. To each reaction is added only one of the four dideoxynucleotides (ddATP, ddGTP, ddCTP, or ddTTP) which are the chain-terminating nucleotides, lacking a 3'-OH group required for the formation of a phosphodiester bond between two nucleotides, thus terminating DNA strand extension and resulting in DNA fragments of varying length.

The newly synthesized and labelled DNA fragments are heat denatured, and separated by size (with a resolution of just one nucleotide) by gel electrophoresis on a denaturing polyacrylamide-urea gel with each of the four reactions run in one of four individual lanes (lanes A, T, G, C); the DNA bands are then visualized byautoradiography or UV light, and the DNA sequence can be directly read off the X-ray film or gel image.

In the image on the right, X-ray film was exposed to the gel, and the dark bands correspond to DNA fragments of different lengths. A dark band in a lane indicates a DNA fragment that is the result of chain termination after incorporation of a dideoxynucleotide (ddATP, ddGTP, ddCTP, or ddTTP). The relative positions of the different bands among the four lanes are then used to read (from bottom to top) the DNA sequence.

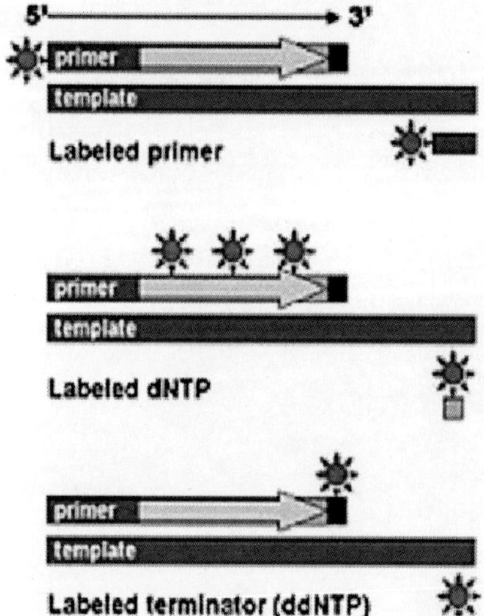

***Figure:** DNA fragments are labelled with a radioactive or fluorescent tag on the primer (1), in the new DNA strand with a labeled dNTP, or with a labeled ddNTP.*

Technical variations of chain-termination sequencing include tagging with nucleotides containing radioactive phosphorus forradiolabelling, or using a primer labeled at the 5' end with a fluorescent dye. Dye-primer sequencing facilitates reading in an optical system for faster and more economical analysis and automation. The later development by Leroy Hood and coworkers of fluorescently labeled ddNTPs and primers set the stage for automated, high-throughput DNA sequencing.

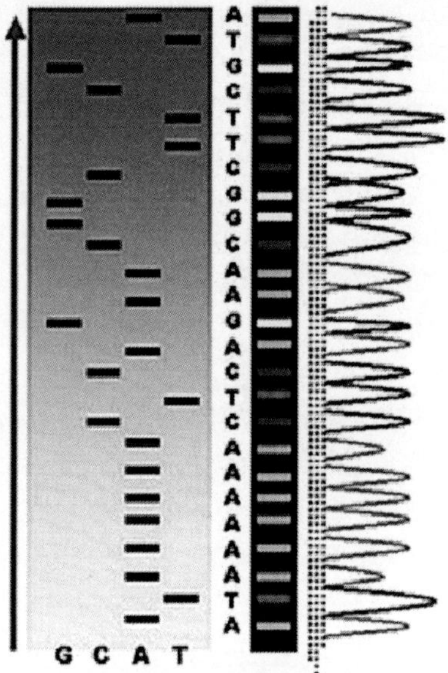

Figure: *Sequence ladder by radioactive sequencing compared to fluorescent peaks.*

Chain-termination methods have greatly simplified DNA sequencing. For example, chain-termination-based kits are commercially available that contain the reagents needed for sequencing, pre-aliquoted and ready to use. Limitations include non-specific binding of the primer to the DNA, affecting accurate read-out of the DNA sequence, and DNA secondary structures affecting the fidelity of the sequence.

Dye-terminator Sequencing

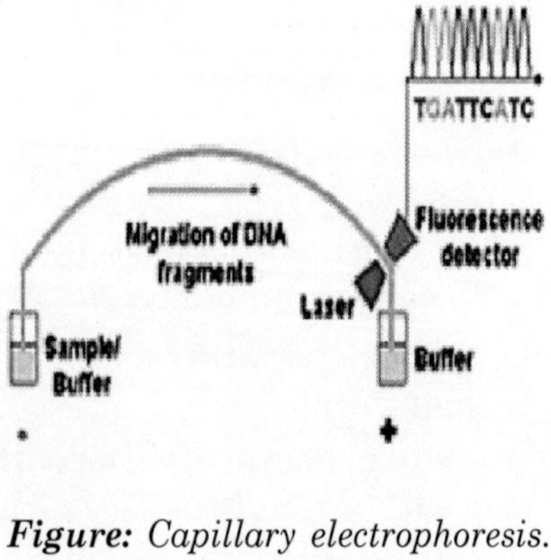

Figure: *Capillary electrophoresis.*

Dye-terminator sequencing utilizes labelling of the chain terminator ddNTPs, which permits sequencing in a single reaction, rather than four reactions as in the labelled-primer method. In dye-terminator sequencing, each of the four dideoxynucleotide chain terminators is labelled with fluorescent dyes, each of which emit light at differentwavelengths.

Owing to its greater expediency and speed, dye-terminator sequencing is now the mainstay in automated sequencing. Its limitations include dye effects due to differences in the incorporation of the dye-labelled chain terminators into the DNA fragment, resulting in unequal peak heights and shapes in the electronic DNA sequence trace chromatogram after capillary electrophoresis.

This problem has been addressed with the use of modified DNA polymerase enzyme systems and dyes that minimize incorporation variability, as well as methods for eliminating "dye blobs". The dye-terminator sequencing method, along with automated high-throughput DNA sequence analyzers, is now being used for the vast majority of sequencing projects.

Challenges

Common challenges of DNA sequencing include poor quality in the first 15–40 bases of the sequence and deteriorating quality of sequencing traces after 700–900 bases. Base calling software typically gives an estimate of quality to aid in quality trimming.

In cases where DNA fragments are cloned before sequencing, the resulting sequence may contain parts of the cloning vector. In contrast, PCR-based cloning and emerging sequencing technologies based on pyrosequencing often avoid using cloning vectors. Recently, one-step Sanger sequencing (combined amplification and sequencing) methods such as Ampliseq and SeqSharp have been developed that allow rapid sequencing of target genes without cloning or prior amplification.

Current methods can directly sequence only relatively short (300–1000 nucleotides long) DNA fragments in a single reaction. The main obstacle to sequencing DNA fragments above this size limit is insufficient power of separation for resolving large DNA fragments that differ in length by only one nucleotide. In all cases the use of a primer with a free 3' end is essential.

Automation and Sample Preparation

Automated DNA-sequencing instruments (DNA sequencers) can sequence up to 384 DNA samples in a single batch (run) in up to 24 runs a day. DNA sequencers carry out capillary electrophoresis for size separation, detection and recording of dye fluorescence, and data output as fluorescent peak tracechromatograms.

Sequencing reactions by thermocycling, cleanup and re-suspension in a buffer solution before loading onto the sequencer are performed separately.

A number of commercial and non-commercial software packages can trim low-quality DNA traces automatically. These programs score the quality of each peak and remove low-quality base peaks (generally located at the ends of the sequence). The accuracy of such algorithms is below visual examination by a human operator, but sufficient for automated processing of large sequence data sets.

Amplification and Clonal Selection

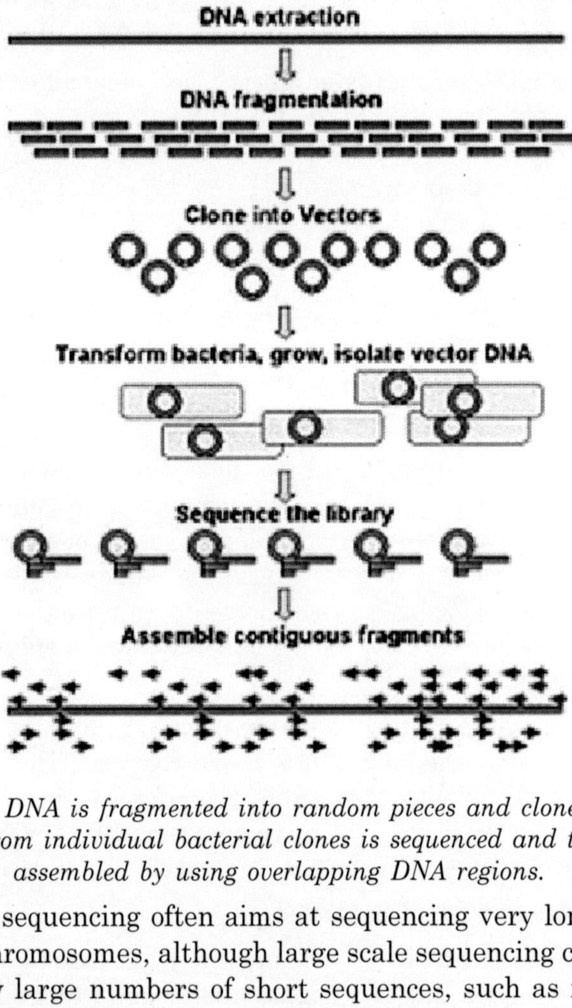

Figure: Genomic DNA is fragmented into random pieces and cloned as a bacterial library. DNA from individual bacterial clones is sequenced and the sequence is assembled by using overlapping DNA regions.

Large-scale sequencing often aims at sequencing very long DNA pieces, such as whole chromosomes, although large scale sequencing can also be used to generate very large numbers of short sequences, such as found in phage display. For longer targets, such as chromosomes, common approaches consist of cutting (with restriction enzymes) or shearing (with mechanical forces) large DNA fragments into shorter DNA fragments.

The fragmented DNA is cloned into aDNA vector, and amplified in Escherichia coli. Short DNA fragments purified from individual bacterial colonies are individually sequenced and assembled electronically into one long, contiguous sequence.

This method does not require any pre-existing information about the sequence of the DNA and is referred to as de novo sequencing. Gaps in the assembled sequence may be filled by primer walking. The different strategies have different tradeoffs in speed and accuracy; shotgun methods are often used for sequencing large genomes, but its assembly is complex and difficult, particularly with sequence repeats often causing gaps in genome assembly.

Most sequencing approaches use an in vitro cloning step to amplify individual DNA molecules, because their molecular detection methods are not sensitive enough for single molecule sequencing. Emulsion PCR isolates individual DNA molecules along with primer-coated beads in aqueous droplets within an oil phase.

Polymerase chain reaction (PCR) then coats each bead with clonal copies of the DNA molecule followed by immobilization for later sequencing. Emulsion PCR is used in the methods by Marguilis et al. (commercialized by 454 Life Sciences), Shendure and Porreca et al. (also known as "Polony sequencing") and SOLiD sequencing, (developed by Agencourt, now Applied Biosystems).

Another method for in vitro clonal amplification is bridge PCR, where fragments are amplified upon primers attached to a solid surface, used in the IlluminaGenome Analyzer. Single-molecule methods, such as that developed by Stephen Quake's laboratory (later commercialized by Helicos) is an exception:

> *it uses bright fluorophores and laser excitation to detect base addition events from individual DNA molecules fixed to a surface, eliminating the need for molecular amplification.*

High-throughput Sequencing

The high demand for low-cost sequencing has driven the development of high-throughput sequencing technologies that parallelize the sequencing process, producing thousands or millions of sequences at once. High-throughput sequencing technologies are intended to lower the cost of DNA sequencing beyond what is possible with standard dye-terminator methods.

The first of the "next-generation" sequencing technologies, MPSS was developed in the 1990s at Lynx Therapeutics, a company founded in 1992 by Sydney Brenner and Sam Eletr. MPSS was a bead-based method that used a complex approach of adapter ligation followed by adapter decoding, reading the sequence in increments of four nucleotides; this method made it susceptible to sequence-specific bias or loss of specific sequences.

Because the technology was so complex, MPSS was only performed 'in-house' by Lynx Therapeutics and no machines were sold; when the merger with Solexa later led to the development of sequencing-by-synthesis, a more simple approach with numerous advantages, MPSS became obsolete. However, the essential properties of the MPSS output were typical of later "next-gen" data types, including hundreds of thousands of short DNA sequences. In the case of MPSS, these were typically used for sequencing cDNA for measurements of gene expression levels. Lynx Therapeutics merged with Solexa in 2004, and this company was later purchased by Illumina.

Polony Sequencing

Polony sequencing, developed in the laboratory of George Church at Harvard, was among the first next-generation sequencing systems used to sequence a full genome in 2005.

It combined an in vitro paired-tag library with emulsion PCR, an automated microscope, and ligation-based sequencing chemistry to sequence an E. coli genome at an accuracy of > 99.9999% and a cost approximately 1/10 that of Sanger sequencing. The technology was licensed to Agencourt Biosciences, subsequently spun out into Agencourt Personal Genomics, and ultimately incorporated into the Applied Biosystems SOLiD platform.

Pyrosequencing

A parallelized version of pyrosequencing was developed by 454 Life Sciences, which has since been acquired by Roche Diagnostics. The method amplifies DNA inside water droplets in an oil solution (emulsion PCR), with each droplet containing a single DNA template attached to a single primer-coated bead that then forms a clonal colony.

The sequencing machine contains many picolitre-volume wells each containing a single bead and sequencing enzymes. Pyrosequencing uses luciferase to generate light for detection of the individual nucleotides added to the nascent DNA, and the combined data are used to generate sequence read-outs. This technology provides intermediate read length and price per base compared to Sanger sequencing on one end and Solexa and SOLiD on the other.

Illumina (Solexa) Sequencing

Solexa, now part of Illumina, developed a sequencing technology based on reversible dye-terminators. DNA molecules are first attached to primers on a slide and amplified so that local clonal colonies are formed (bridge amplification).

Four types of ddNTPs are added, and non-incorporated nucleotides are washed away. Unlike pyrosequencing, the DNA can only be extended one nucleotide at a time. A camera takes images of the fluorescently labeled nucleotides, then the dye along with the terminal 3' blocker is chemically removed from the DNA, allowing the next cycle.

Solid Sequencing

Applied Biosystems' SOLiD technology employs sequencing by ligation. Here, a pool of all possible oligonucleotides of a fixed length are labeled according to the sequenced position. Oligonucleotides are annealed and ligated; the preferential ligation by DNA ligase for matching sequences results in a signal informative of the nucleotide at that position. Before sequencing, the DNA is amplified by emulsion PCR. The resulting bead, each containing only copies of the same DNA molecule, are deposited on a glass slide. The result is sequences of quantities and lengths comparable to Illumina sequencing.

Ion Semiconductor Sequencing

Ion Torrent Systems Inc. developed a system based on using standard sequencing chemistry, but with a novel, semiconductor based detection system. This method of sequencing is based on the detection of hydrogen ions that are released during the polymerisation of DNA, as opposed to the optical methods used in other sequencing systems. A microwell containing a template DNA strand to be sequenced is flooded with a single type of nucleotide. If the introduced nucleotide is complementary to the leading template nucleotide it is incorporated into the growing complementary strand.

This causes the release of a hydrogen ion that triggers a hypersensitive ion sensor, which indicates that a reaction has occurred. If homopolymer repeats are present in the template sequence multiple nucleotides will be incorporated in a single cycle. This leads to a corresponding number of released hydrogens and a proportionally higher electronic signal.

DNA Nanoball Sequencing

DNA nanoball sequencing is a type of high throughput sequencing technology used to determine the entire genomic sequence of an organism. The company Complete Genomics uses this technology to sequence samples that researchers submit from several projects. The method uses rolling circle replication to amplify small fragments of genomic DNA into DNA nanoballs. Unchained sequencing by ligation is then used to determine the nucleotide sequence. This method of DNA sequencing allows large numbers of DNA nanoballs to be sequenced per run and at low reagent costs compared to other next generation sequencing platforms.

However, only short sequences of DNA are determined from each DNA nanoball which makes mapping the short reads to a reference genome difficult. This technology has been used for multiple genome sequencing projects and is scheduled to be used for more.

Helioscope(TM) Single Molecule Sequencing

Based on "true single molecule sequencing" technology, Helioscope sequencing uses DNA fragments with added polyA tail adapters, which are attached to the flow cell surface. The next steps involve extension-based sequencing with cyclic washes of the flow cell with fluorescently labeled nucleotides (one nucleotide type at a time, as with the Sanger method).

The reads are performed by the Helioscope sequencer. The reads are short, up to 55 bases per run, but recent improvemend of the methodology allowes more accurate reads of homopolymers (stretches of one type of nucleotides) and RNA sequencing.

Single Molecule SMRT(TM) Sequencing

SMRT sequencing is based on the sequencing by synthesis approach. The DNA is synthesisd in so calles zero-mode wave-guides (ZMWs) - small well-like containers with the capturing tools located at the bottom of the well. The sequencing is performed with use of unmodified polymerase (attached to the ZMW bottom) and fluorescently labelled nucleotides flowing freely in the solution. The wells are constructed in a way that only the fluorescence occurring by the bottom of the well is detected. The fluorescent label is detached from the nucleotide at its incorporation into the DNA strand, leaving an unmodified DNA strand. According to Pacific Biosciences, the SMTR technology developer, this methodology allows detection of nucleotide modifications (such ad cytosine methylation). This happens through the observation of polymerase kinetics. This approach allows reads of 1000 nucleotides.

Single Molecule Real Time (RNAP) Sequencing

This method is based on RNA polymerase (RNAP), which is attached to a polystyrene bead, with distal end of sequenced DNA is attached to another bead, with both beads being placed in optical traps. RNAP motion during transcription brings the beads in closer and their relative distance changes, which can then be recorded at a single nucleotide resolution. The sequence is deduced based on the four readouts with lowered concentrations of each of the four nucleotide types.

Nanopore DNA Sequencing

This method is based on the readout of electrical signal occurring at nucleotides passing by alpha-hemolysin pores covalently bound with cyclodextrin. The DNA passing through the nanopore changes its ion current. This change is dependent on the shape, size and length of the DNA sequence. Each type of the nucleotide blocks the ion flow through the pore for a different period of time. The method has a potential of development as it does not require modified nucleotides, however single nucleotide resolution is not yet available.

VisiGen Biotechnologies Approach

VisiGen Biotechnologies introduced a specially engineered DNA polymerase for use in their sequencing. This polymerase acts as a sensor - having incorporated a donor fluorescent dye by its active centre. This donor dye acts by FRET (fluorescent resonant energy transfer), inducing fluorescence of differently labeled nucleotides. This approach allows reads performed at the spead at which polymerase incorporates nucleotides into the sequence (several hundred per second). The nucleotide fluorochrome is released after the incorporation into the DNA strand. The expected read lengths in this approach should reach 1000 nucleotides, however this will have to be confirmed.

Future Methods

Sequencing by hybridization is a non-enzymatic method that uses a DNA microarray. A single pool of DNA whose sequence is to be determined is fluorescently labeled and hybridized to an array containing known sequences.

Strong hybridization signals from a given spot on the array identifies its sequence in the DNA being sequenced. Mass spectrometry may be used to determine mass differences between DNA fragments produced in chain-termination reactions. DNA sequencing methods currently under development include labeling the DNA polymerase, reading the sequence as a DNA strand transits through nanopores, and microscopy-based techniques, such as AFM or transmission electron microscopy that are used to identify the positions of individual nucleotides within long DNA fragments (>5,000 bp) by nucleotide labeling with heavier elements (e.g., halogens) for visual detection and recording.

Third generation technologies aim to increase throughput and decrease the time to result and cost by eliminating the need for excessive reagents and harnessing the processivity of DNA polymerase. In microfluidic Sanger sequencing the entire thermocycling amplification of DNA fragments as well as their separation by electrophoresis is done on a single glass wafer (approximately 10 cm in diameter) thus reducing the reagent usage as well as cost. In some instances researchers have shown that they can increase the throughput of conventional sequencing through the use of microchips. Research will still need to be done in order to make this use of technology effective.

In October 2006, the X Prize Foundation established an initiative to promote the development of full genome sequencing technologies, called the Archon X Prize, intending to award $10 million to "the first Team that can build a device and use it to sequence 100 human genomes within 10 days or less, with an accuracy of no more than one error in every 100,000 bases sequenced, with sequences accurately covering at least 98% of the genome, and at a recurring cost of no more than $10,000 (US) per genome."

Each year NHGRI promotes grants for new research and developments in genomics. 2010 grants and 2011 candidates include continuing work in microfluidic, polony and base-heavy sequencing methodologies

Major Landmarks in DNA Sequencing

- 1953 Discovery of the structure of the DNA double helix.
- 1972 Development of recombinant DNA technology, which permits isolation of defined fragments of DNA; prior to this, the only accessible samples for sequencing were from bacteriophage or virus DNA.
- 1977 The first complete DNA genome to be sequenced is that of bacteriophage öX174.
- 1977 Allan Maxam and Walter Gilbert publish "DNA sequencing by chemical degradation". Frederick Sanger, independently, publishes "DNA sequencing with chain-terminating inhibitors".
- 1984 Medical Research Council scientists decipher the complete DNA sequence of the Epstein-Barr virus, 170 kb.
- 1986 Leroy E. Hood's laboratory at the California Institute of Technology and Smith announce the first semi-automated DNA sequencing machine.
- 1987 Applied Biosystems markets first automated sequencing machine, the model ABI 370.
- 1990 The U.S. National Institutes of Health (NIH) begins large-scale sequencing trials on Mycoplasma capricolum, Escherichia coli, Caenorhabditis elegans, and Saccharomyces cerevisiae (at US$0.75/ base).
- 1991 Sequencing of human expressed sequence tags begins in Craig Venter's lab, an attempt to capture the coding fraction of the human genome.
- 1995 Craig Venter, Hamilton Smith, and colleagues at The Institute for Genomic Research (TIGR) publish the first complete genome of a free-living organism, the bacterium Haemophilus influenzae. The circular chromosome contains 1,830,137 bases and its publication in the journal Science marks the first use of whole-genome shotgun sequencing, eliminating the need for initial mapping efforts.
- 1996 Pål Nyrén and his student Mostafa Ronaghi at the Royal Institute of Technology in Stockholm publish their method of pyrosequencing
- 1998 Phil Green and Brent Ewing of the University of Washington publish "phred" for sequencer data analysis.
- 2000 Lynx Therapeutics publishes and markets "MPSS" - a parallelized, adapter/ligation-mediated, bead-based sequencing technology, launching "next-generation" sequencing.
- 2001 A draft sequence of the human genome is published.

- 2004 454 Life Sciences markets a parallelized version of pyrosequencing. The first version of their machine reduced sequencing costs 6-fold compared to automated Sanger sequencing, and was the second of a new generation of sequencing technologies, after MPSS.

Complete Genomics

Complete Genomics is a life sciences company that has developed and commercialized a DNA sequencing platform for human genome sequencing and analysis. This solution combines the company's proprietary human genome sequencing technology with its informatics and data management software in an outsourced service model that provides its customers with finished variant reports and assemblies. Researchers send DNA samples to the company, which are sequenced at Complete Genomics' own commercial genome center in Mountain View, California. Sequencing is completed using the company's proprietary instruments and technology. Assembled sequences and variant reports are then delivered back to the researcher.

History

Complete Genomics was founded in March 2006 by Clifford Reid, Radoje (Rade) Drmanac, and John Curson. Clifford Reid is the chairman, president and chief executive officer of Complete Genomics.

In Feb. 2009, Complete Genomics announced that it had sequenced its first human genome and submitted the resulting variant data to the National Center for Biotechnology Information database. Then in Nov. 2009, Complete Genomics published sequence data for three human genomes in the journal Science.

By the end of that year, Complete Genomics had sequenced 50 human genomes. The company has now significantly increased the throughput in its genome sequencing factory; it sequenced and analyzed 300 complete human genomes in Q3 2010.

In addition, Complete Genomics' customers are now publishing papers describing research completed using data from Complete Genomics. Examples include the Institute for Systems Biology's project to sequence a family of four and verify the gene responsible for Miller Syndrome, a rare craniofacial disorder and Genentech's work to sequence and compare a patient's primary lung tumor and adjacent normal tissue.

Technology Platform

Complete Genomics' proprietary human genome sequencing technology is optimized for the exclusive study of human DNA, providing assembled sequences and variation files. The technology relies onDNA nanoball sequencing. It is designed to use lower volumes and concentrations of reagents than existing systems. Additionally, Complete Genomics' proprietary instrumentation allows for greater numbers of base reads per image.

DNA Field-effect Transistor

A DNA field-effect transistor (DNAFET) is a field-effect transistor which uses the field-effect due to the partial charges of DNA molecules to function as a biosensor. The structure of DNAFETs is similar to that of MOSFETs with the exception of the gate structure which, in DNAFETs, is replaced by a layer of immobilized ssDNA (single-stranded DNA) molecules which act as surface receptors. When complementary DNA strands hybridize to the receptors, the charge distribution near the surface changes, which in turn modulates current transport through the semiconductor transducer.

Arrays of DNAFETs can be used for detecting single nucleotide polymorphisms (causing many hereditary diseases) and for DNA sequencing. Their main advantage compared to optical detection methods in common use today is that they do not require labeling of molecules.

DNA Sequencing Theory

DNA sequencing theory is the broad body of work that attempts to lay analytical foundations for DNA sequencing. The practical aspects revolve around designing and optimizing sequencing projects (known as "strategic genomics"), predicting project performance, troubleshooting experimental results, characterizing factors such as sequence bias and the effects of software processing algorithms, and comparing various sequencing methods to one another. In this sense, it could be considered a branch of systems engineering or operations research.

The permanent archive of work is primarily mathematical, although numerical calculations are often conducted for particular problems too. DNA sequencing theory addresses physical processes related to sequencing DNA and should not be confused with theories of analyzing resultant DNA sequences, e.g. sequence alignment. Publications sometimes do not make a careful distinction, but the latter are primarily concerned withalgorithmic issues.

Sequencing as a Covering Problem

All mainstream methods of DNA sequencing rely on reading small fragments of DNA and subsequently reconstructing these data to infer the original DNA target, either via assembly or alignment to a reference. The abstraction common to these methods is that of a mathematical covering problem.

For example, one can imagine a line segment representing the target and a subsequent process where smaller segments are "dropped" onto random locations of the target. The target is considered "sequenced" when adequate coverage accumulates, for example when no gaps remain. The abstract properties of covering have been studied by mathematicians for over a century. However, direct application of these results has not generally been possible. Closed-form mathematical solutions, especially for probability distributions, often cannot be readily evaluated.

That is, they involve inordinately large amounts of computer time for parameters characteristic of DNA sequencing. Stevens' configuration is one such example. Results obtained from the perspective of pure mathematics also do not account for factors that are actually important in sequencing, for instance detectable overlap in sequencing fragments, double-stranding, edge-effects, and target multiplicity.

Consequently, development of sequencing theory has proceeded more according to the philosophy ofapplied mathematics. In particular, it has been problem-focused and makes expedient use of approximations, simulations, etc.

Early Uses Derived from Elementary Probability Theory

The earliest result was actually borrowed directly from elementary probability theory. If we model the above process and take L and G as the fragment length and target length, respectively, then the probability of "covering" any given location on the target with one particular fragment is L / G.

Note that this presumes $L \ll G$, which is valid for many, though not all sequencing scenarios. Utilizing concepts from the binomial distribution, it can then be shown that the probability that the location is covered by at least one of N fragments is

$$P = 1 - \left[1 - \frac{L}{G}\right]^N .$$

This equation was first used to characterize plasmid libraries, but is often more useful in a modified form. For most projects , so that, to a good degree of approximation

$$\left[1 - \frac{L}{G}\right]^N \sim \exp\left(-NL/G\right),$$

where R = NL / G is called the redundancy. Note the significance of redundancy as representing the average number of times a position is covered with fragments. Note also that in considering the covering process over all positions in the target, this probability is identical to the expected value of the random variable C, which represents the fraction of the target coverage. The final result,

$$E\langle C\rangle = 1 - e^{-R},$$

remains in widespread use as a "back of the envelope" estimator and predicts that coverage for all projects evolves along a universal curve that is a function only of the redundancy.

Lander-Waterman Theory

In 1988, Eric Lander and Michael Waterman published an important paper examining the covering problem from the standpoint of gaps. Although they focused on the so-called mapping problem, the abstraction to sequencing is much the same.

They furnished a number of useful results that were adopted as the standard theory from the earliest days of "large-scale" genome sequencing. Their model was also used in designing the Human Genome Project and continues to play an important role in DNA sequencing.

Ultimately, the main goal of a sequencing project is to close all gaps, so the "gap perspective" was a logical basis of developing a sequencing model. One of the more frequently used results from this model is the expected number of contigs, given the number of fragments sequenced. If one neglects the amount of sequence that is essentially "wasted" by having to detect overlaps, their theory yields

$$E\langle contings\rangle = Ne^{-R}.$$

In 1995, Roach published improvements to this theory, enabling it to be applied to sequencing projects in which the goal was to completely sequence a target genome. Wendl and Waterstonconfirmed, based on Stevens' method, that both models produced similar results when the number of contigs was substantial, such as in low coverage mapping or sequencing projects.

As sequencing projects ramped up in the 1990s, and projects approached completion, low coverage approximations became inadequate, and the exact model of Roach was necessary. However, as the cost of sequencing dropped, parameters of sequencing projects became easier to directly test empirically, and interest and funding for strategic genomics diminished

The basic ideas of Lander-Waterman theory led to a number of additional results for particular variations in mapping techniques. However, technological advancements have rendered mapping theories largely obsolete except in organisms other than highly studied model organisms (e.g., yeast, flies, mice, and humans).

Parking Strategy

The parking strategy for sequencing resembles the process of parking cars along a curb. Each car is a sequenced clone, and the curb is the genomic target. Each clone sequenced is screened to ensure that subsequently sequenced clones do not overlap any previously sequenced clone. No sequencing effort is redundant in this strategy. However, much like the gaps between parked cars, unsequenced gaps less than the length of a clone accumulate between sequenced clones. There can be considerable cost to close such gaps.

Pairwise End-sequencing

In 1995, Roach et al. proposed and demonstrated through simulations a generalization of a set of strategies explored earlier by Edwards and Caskey. This whole-genome sequencing method became immensely popular as it was championed by Celera and used to sequenced several model organisms before Celera applied it to the human genome. Today, most sequencing projects employ this strategy, often called paired end sequencing.

Recent Advancements

The physical processes and protocols of DNA sequencing have continued to evolve, largely driven by advancements in bio-chemical methods, hardware, and automation techniques. There is now a wide range of problems that DNA sequencing has made in-roads into, including metagenomics and medical (cancer) sequencing. There are important factors in these scenarios that classical theory does not account for. Recent work has begun to focus on resolving the effects of some of these issues. The level of mathematics becomes commensurately more sophisticated.

Multiplicity

Biologists have developed methods to filter highly-repetitive, essentially un-sequenceable regions of genomes. These procedures are important for organisms whose genomes consist mostly of such DNA, for example corn. They yield multitudes of small islands of sequenceable DNA products. Wendl and Barbazuk proposed an extension to Lander-Waterman Theory to account for "gaps" in the target due to filtering and the so-called "edge-effect".

The latter is a position-specific sampling bias, for example the terminal base position has only a 1 / G chance of being covered, as opposed toL / G for interior positions. For R < 1, classical Lander-Waterman Theory still gives good predictions, but dynamics change for higher redundancies.

Small Versus Large Fragments

Modern sequencing methods usually sequence both ends of a larger fragment, which provides linking information for de novo assembly and improved probabilities for alignment to reference sequence. Researchers generally believe that longer lengths of data (read lengths) enhance performance for very large DNA targets, an idea consistent with predictions from distribution models.

However, Wendl showed that smaller fragments provide better coverage on small, linear targets because they reduce the edge effect in linear molecules. These findings have implications for sequencing the products of DNA filtering procedures. Read-pairing and fragment size evidently have negligible influence for large, whole-genome class targets.

Diploid Sequencing

Sequencing is emerging as an important tool in medicine, for example in cancer research. Here, the ability to detect heterozygous mutations is important and this can only be done if the sequence of the diploid genome is obtained. In the pioneering efforts to sequence individuals, Levy et al. and Wheeler et al., who sequenced Craig Venter and Jim Watson, respectively, outlined models for covering both alleles in a genome.

Wendl and Wilson followed with a more general theory that allowed for an arbitrary number of coverings of each allele and arbitrary ploidy. These results point to the general conclusion that the amount of data needed for such projects is significantly higher than for traditional haploid projects.

Limitations

DNA sequencing theories often invoke the assumption that certain random variables in a model are independently and identically distributed. For example, in Lander-Waterman Theory, a sequenced fragment is presumed to have the same probability of covering each region of a genome and all fragments are assumed to be independent of one another.

In actuality, sequencing projects are subject to various types of bias, including differences of how well regions can be cloned, sequencing anomalies, biases in the target sequence (which is not random), and software-dependent errors and biases. In general, theory will agree well with observation up to the point that enough data have been generated to expose latent biases. The kinds of biases related to the underlying target sequence are particularly difficult to model, since the sequence itself may not be known a priori. This presents a type of "chicken and egg" closure problem.

Academic Status

Sequencing theory is based on elements of mathematics, biology, and systems engineering, so it is highly interdisciplinary. Although many universities now have programs in computational biology, there does not yet seem to be a strong focus at the graduate level on this topic. Academic contributions have mainly been limited to a small number of PhD dissertations.

Bibliography

Annie Tan: *Blueprints Notes and Cases - Biochemistry, Genetics and Embryology*, Blackwell Science, Delhi, 2004.

Balkwill, Fran: *DNA Is Here to Stay*, Carolrhoda Books, Inc., Minneapolis, 1993.

Banerjee, S : *Cell Biochemistry*, Dominant Pub, Delhi, 2009.

Bhatia, A L : *Biochemistry and Endocrinology*, Indus Valley, Delhi, 2002.

Brookes, Martin: *Get a Grip on Genetics*, *Time Life Books*, East Sussex, England, 1998.

Bruce, Donald : *Engineering Genesis: The Ethics of Genetic Engineering in Non-Human Animals*, London: Earthscan, 1998.

Cahill, Lisa: *Genetics, Theology, and Ethics: An Interdisciplinary Conversation*, New York: Crossroad, 2005.

Curtis, M.: *The Geometry of DNA: A Structural Revision*. London: Blue Gallery, 1996.

Diaz, E.: *Microbial Biodegradation: Genomics and Molecular Biology,* Caister Academic Press, UK, 2008.

Drlica, Karl : *Understanding DNA and Gene Cloning: A Guide for the Curious*, New York: Wiley, 1997.

Fukuyama F.: *Our Posthuman Future: Consequences of the Biotechnology Revolution,* Farrar, Straus and Giroux, NewYork, 2002.

Garza-Valdes, L.A.: *The DNA of God?* Hodder and Stoughton, London, 1998.

Hames, B. D.: *Genes and Embryos*, Oxford University Press, New York, 1989.

Jonoska, N.: *Self-Assembling DNA Graphs, DNA-Based Computers VIII*, Springer-Verlag, Berlin, 2003.

Kallenbach, N.R.: *Molecular Structure: Chemical Reactivity and Biological Activity,* Oxford University Press, Oxford, 1988.

Lewis, Ricki: *Human Genetics: Concepts and Applications,* Boston, Mass: McGraw Hill. 1994.

Mayr, E.: *Change of Genetic Environment and Evolution*, Allen and Unwin, London, 1954.

Nisha Khalsa: *Essentials of Biochemistry*, Aavishkar Pub, Delhi, 2008.

Old, R.W.: *Principles of Gene Manipulation*, London, Blackwell Scientific Publications, 1989.

Rabus, R: *Microbial Biodegradation: Genomics and Molecular Biology*, Caister Academic Press, UK, 2008.

Sturtevant, Alfred: *History of Genetics*, Harper and Row, New York, 1965.

Stwertka, Eve : *Genetic Engineering*, Franklin Watts, New York, 1982.

Stwertka, Eve and Albert: *Genetic Engineering*, Franklin Watts, New York, 1982.

Vasudev Bhatnagar: *Cell Science and Technology*, Campus Books, Delhi, 2009.

William, Kovacs D.: *An Introduction to Geotechnical Engineering*, Prentice Hall, UK, 1981.

Williams, J.G. and R.K. Patient : *Genetic Engineering*. Oxford, England: IRL Press, 1988.

Wilson, I.: *The Blood and The Shroud*. London: Weidenfeld & Nicolson, 1998.

Winston, R.: *The Future of Genetic Manipulation*, Phoenix, London, 1997.

Index

❑❑❑